The Emergence of Consciousness

edited by

Anthony Freeman

IMPRINT ACADEMIC

Cover Illustration

Brain and Mind (detail) © Herms Romijn

Herms Romijn, who painted the water-colour picture of which a detail is reproduced on the front cover, has exhibited widely and sold many paintings in his native Holland. A neurobiologist by profession, he has published papers on the pineal gland, neuron culture, epilepsy and the biological clock, and is the author of books (in Dutch) on sleep and mind–brain relationships. In recent paintings, he has tried to express the essence of his view on consciousness, which were also described in a recent monograph. He sees consciousness as a manifestation of complex patterns of electric and/or magnetic fields in the brain, pointing out that virtual photons comprising these fields can therefore, in a sense, be regarded as 'elementary carriers' of consciousness. Because not only fields but neurons too are composed of elementary particles, he used the technique of pointillism to emphasize this feature in the painting.

Published in the UK by Imprint Academic
PO Box 1, Thorverton EX5 5YX, UK

Published in the USA by Imprint Academic
Philosophy Documentation Center, PO Box 7147, Charlottesville, VA 22906-7147, USA

World Copyright © Imprint Academic, 2001
No part of any contribution may be reproduced in any form without permission,
except for the quotation of brief passages in criticism and discussion.
The opinions expressed in the articles and book reviews are
not necessarily those of the editors or the publishers.

ISBN 0907845 18 5

British Library Cataloguing in Publication data
A catalogue record for this book is available from the British Library
Library of Congress Card Number 2001094506

ISSN: 1355 8250 (*Journal of Consciousness Studies*, **8**, No. 9–10, 2001)

JCS is indexed and abstracted in
Social Sciences Citation Index[®], ISI Alerting Services (includes *Research Alert*[®]),
Current Contents[®]: *Social and Behavioral Sciences, Arts and Humanities Citation Index*[®],
Current Contents[®]: *Arts & Humanities Citation Index*[®], *Social Scisearch*[®], *PsycINFO*[®]
and *The Philosopher's Index*.

Journal of Consciousness Studies
controversies in science & the humanities

Vol. 8, No. 9–10, September/October 2001

SPECIAL ISSUE: 'THE EMERGENCE OF CONSCIOUSNESS'

Edited by Anthony Freeman

TABLE OF CONTENTS

iv	About Authors	
vii	Editorial Introduction	*Anthony Freeman*
1	Reduction, Emergence and Other Recent Options on the Mind–Body Problem: A Philosophic Overview	*Robert Van Gulick*
35	Some Perils of Quantum Consciousness: Epistemological Pan-experientialism and the Emergence–Submergence of Consciousness	*Harry T. Hunt*
47	Emergence and the Uniqueness of Consciousness	*Natika Newton*
61	Converging on Emergence: Consciousness, Causation and Explanation	*Michael Silberstein*
99	Constraints on an Emergent Formulation of Conscious Mental States	*Scott Hagan and Masayuki Hirafuji*
123	Why the Mind Is Not a Radically Emergent Feature of the Brain	*Todd E. Feinberg*
147	God as an Emergent Property	*Anthony Freeman*
161	We Could Be Siblings Yet: Reflections on Huston Smith's *Why Religion Matters*	*Alwyn Scott*

ABOUT AUTHORS

Todd E. Feinberg (Beth Israel Medical Center, Fierman Hall, 317 East 17th Street, New York, NY 10003, USA) is chief of the Yarmon Neurobehavior and Alzheimer's Disease Center, Beth Israel Medical Center in New York City, and associate professor of neurology and psychiatry, Albert Einstein College of Medicine. He is co-editor (with Martha J. Farah) of the textbook *Behavioral Neurology and Neuropsychology* (McGraw-Hill, 1997) and author of the recently published book *Altered Egos: How the Brain Creates the Self* (Oxford University Press, 2001) from which his article in this volume is adapted.

Anthony Freeman (Imprint Academic, PO Box 1, Thorverton, Devon EX5 5YX, UK) took degrees in chemistry and then theology at Oxford University. He was ordained in 1972 and held a variety of pastoral and teaching posts in the Church of England until being dismissed for publishing the allegedly heretical book *God In Us* in 1993 (second edition 2001). He remains a priest and lectures and writes on theology and consciousness matters. He has been managing editor of the *Journal of Consciousness Studies* since its launch in 1994. His other published work includes the books *Gospel Treasure* (1999) and *The Volitional Brain* (co-edited 1999).

Robert Van Gulick (Department of Philosophy, 541 HL, Syracuse University, Syracuse, NY 13244-1170, USA) is professor of philosophy at the University of Syracuse and the director of the University's Cognitive Science Program. His work has focused on the philosophy of psychology, including such topics as mental representation, intentional content, self-consciousness, and mental causation. He has numerous publications and presentations on these and related topics and is a co-editor of *John Searle and his Critics* (Blackwell,1990).

Scott Hagan (Mathematics Department, British Columbia Institute of Technology, 3700 Willingdon Avenue, Burnaby, British Columbia, V5G 3H2 Canada) received his PhD in high-energy physics and cosmology from McGill University in 1995. His research has explored topics in high-energy physics (scale invariance, supergravity, wormholes), biophysics (cellular modelling) and the physical and philosophical issues surrounding quantum and classical models of consciousness. He has recently been working in Japan, modelling quantum effects in biology, and now teaches mathematics at the British Columbia Institute of Technology in Vancouver, Canada.

Harry T. Hunt (Department of Pychology, Brock University, St Catherine's, Ontario, L2S 3A1 Canada) received his PhD from Brandeis University in 1971. He is professor of psychology at Brock University, St Catherine's, Ontario, and adjunct professor at the California Institute for Human Science, San Diego. He is the author of *The Multiplicity of Dreams* (1989) and *On the Nature of Consciousness*(1995), both with Yale University Press. His interests lie at the intersections of cognitive, phenomenological, transpersonal, psychodynamic, and sociological approaches to consciousness. He has published empirical studies on lucid dreaming, dream bizarreness, meditative states, creativity and metaphor, and transpersonal experiences in childhoood and theoretical papers on mystical experience based on a cognitive psychology of microgenesis and synesthesia.

ABOUT AUTHORS

Masayuki Hirafuji (Computational Modeling Lab, Department of Information Research, NARC, 3-1-1 Kannondai, Tsukuba, Ibaraki, 305-8666 Japan) obtained his PhD in environmental/computational biology in 1981 from the University of Tokyo. He was appointed Research and National Project Coordinator for the Japanese government from 1994 to 1996. Since 1985 he has been conducting research at the Computational Modeling Lab at NARC in Tsukuba, Japan, where he is investigating computational modelling methods for biological systems. He is especially interested in the fundamental mechanisms common to an understanding of consciousness, life and evolution.

Natika Newton (Department of Philosophy, Nassau County Community College, Garden City, NY 11530, USA) received her PhD from the State University of New York at Stoney Brook in 1981. She has been developing theories of consciousness and intentionality for the past twenty years, publishing papers in many journals. Her book *Foundations of Understanding* (Benjamins, 1996) argues that all levels of intentionality are rooted in conscious experiences of embodied action. She is founding co-editor with Ralph Ellis of the journal *Consciousness and Emotion*, launched by Benjamins in 2000.

Michael Silberstein (Department of Philosophy, Wenger Center, Elizabethtown College, One Alpha Drive, Elizabethtown, PA 17022, USA) is associate professor of philosophy at Elizabethtown College. He is an NEH fellow who has published and delivered papers on both philosophy of science and philosophy of mind. His primary research interests are philosophy of physics and of cognitive neuroscience. He is especially interested in how these branches of philosophy bear on more general questions of reduction, emergence and explanation. For more on these topics see his forthcoming (January, 2002) co-edited volume with Peter Machamer, *Blackwell Guide to the Philosophy of Science* (chapter five).

Alwyn Scott (Institute of Mathematical Modelling, Technical University of Denmark, DK-2800 Lyngby, Denmark) is an emeritus professor of mathematics at the University of Arizona, a professor in the Department of Informatics and Mathematical Modelling at the Technical University of Denmark, and the author of *Stairway to the Mind,* published by Copernicus in 1995. He is currently working on a book entitled *Neuroscience: A Mathematical Primer,* to be published by Springer-Verlag, New York in the spring of 2002.

Anthony Freeman

Editorial Introduction

This set of essays came as a gift. Such collections normally result from invitations, either to contribute directly to a book or to take part in a workshop or conference out of which a publication proceeds, but in this instance things worked otherwise. The contributions all came in as unsolicited submissions to the *Journal of Consciousness Studies*, their arrival spread over a period of some months. Only at some point during the review process did the individual papers present themselves to me as a group, as potential parts of a special issue on the topic of emergence. Rather as the Dalmatian dog or the face of Christ reveals itself from a random array of splodges in the familiar gestalt 'pictures', so did this collection stand out and offer itself to me — unsought — from among the hundred or so papers then under review. You might say that it 'emerged'.

The collection's gift-like quality has not, however, meant that the volume's production has been pain-free. Almost without exception, the authors have toiled under the scrutiny of the journal's referees and editors and revised their work more than once in bringing it to its present state. I am grateful to all of the contributors for their forbearance and co-operation, and wish especially to thank Joseph Goguen, editor-in-chief of the *Journal of Consciousness Studies*, for the special interest he has taken in this particular venture. I am also conscious of my debt to Robert Van Gulick. After 'Tucson 2000' he asked whether the journal might be interested in a paper written up from the notes for his pre-conference workshop on emergence. The offer was accepted, and I have no doubt it was having the promise of this paper in the back of my mind that predisposed me to 'see' the complete collection when it presented itself.

For this reason, as well as the value of starting a volume such as this with an overview of its subject matter, **Robert Van Gulick**'s 'Reduction, Emergence, and Other Recent Options on the Mind–Body Problem' has been given pride of place at the head of the contributions. The paper sets out 'family trees' showing how the various kinds of reduction and emergence parallel and relate to each other. The taxonomy carefully distinguishes between theories that concern 'real' underlying ontological/metaphysical relations and those that restrict themselves to representational/epistemic questions that arise from our human outlook on the world. As promised in his title, he also considers how some 'other recent options' fit into his pattern of theories on the mind–body problem. He concludes — somewhat surprisingly — that the 'logical space' allocated to non-reductive physicalism is not additional to that occupied by emergence and reduction, but is rather 'a special sub-region' where these opposed tendencies intersect.

The rival claims of ontology and epistemology in discussions of emergence are further explored by **Harry T. Hunt** in 'Some Perils of Quantum Consciousness'. He acknowledges a similarity between microphysics and phenomenal consciousness, in that both carry us to the limits of observation, but argues that this is not a sound basis on which to make ontological pronouncements concerning the nature of their relation. He sees in consciousness and quantum-level physics two expressions 'on very different levels of complexity' of the same organizing principles, i.e. emergentist ones. Thus a two-way metaphorical relationship is found between phenomenology and microphysics. It is because of these 'intrinsically bi-directional' perceptual structures that the ways the two 'can inform each other epistemologically need to be kept separate . . . from ontological claims of explanation'.

In 'Emergence and the Uniqueness of Consciousness', **Natika Newton** offers what she hopes is 'a happy blend of physicalist explanation with respectful acknowledgement of the robustness of subjective experience'. If that combination sounds a tall order, it is probably no accident. Newton here argues that the need for subjects to cope with 'the forced blending of components that are incompatible' is precisely what gives rise to phenomenal consciousness. This she presents as a special case of a more general observation that 'novelties emerge from incompatibilities'.

Among today's philosophers of mind a leading protagonist of 'radical' emergence is **Michael Silberstein**. In 'Converging on Emergence' he first sets out what he takes to be the three requirements of any theory of consciousness: to take consciousness seriously; to show how it is conceivable/possible that consciousness arises from fundamental elements that are not conscious; and to show how it is conceivable/possible that conscious states can causally interact with neurochemical states. He then argues powerfully for the view that the rivals to radical emergence all fall down on one or more of these criteria, leaving it the most plausible contender.

If the combined advocacy of the previous three papers carries us along too speedily towards the conclusion that emergence is the best or only solution to the mind–body problem, then **Scott Hagan** and **Masayuki Hirafuji** provide a cautionary amber light by drawing attention to some 'Constraints on an Emergent Formulation of Conscious Mental States'. They take the widest possible view of what might constitute an 'emergent formulation', and include both functionalism and computationalism among those theories of mind 'which are ultimately justified by an appeal to emergentist principles'. From here they proceed to ask whether emergent accounts of conscious mental states can fulfill their required aims within the paradigm of purely *classical* science, understood as entailing microphysical determinism. Distinguishing first between local and global states as the basis for emergence, and then between extrinsic and intrinsic modes of description, and finally focusing on the question of locality, the authors conclude that the constraints of classical science do spell trouble for emergent formulations. Hagan and Hirafuji reject a number of unpalatable alternatives, such as those which deny the ontology of conscious states or offend against parsimony, and end up drawn towards some form of quantum solution.

Hagan and Hirafuji's warning amber light seems to be backed up by an uncompromising red one in **Todd Feinberg**'s 'Why the Mind is Not a Radically Emergent Feature of the Brain'. It soon becomes clear, however, that Feinberg is very much in tune with the aims of the emergentists. Like them he makes the case for a physically grounded account of the mind–brain relationship, which nonetheless produces irreducibly personal mental states, but he finds their *radical* emergence an unnecessary postulate. The key to his interpretation is the concept of 'nested hierarchies', a holistic concept in which the various levels are composed of each other, as opposed to non-nested or pyramidal hierarchies in which there is a clear-cut top and bottom. He is thus in agreement with John Searle that consciousness is what Searle calls an 'emergent1 property' (where the higher level is 'caused by and realized in' the lower one) but not what Searle calls an 'emergent2 property'. This latter is the radical version of emergence advocated by Silberstein and denied in Feinberg's title.

The two concluding contributions both consider the emergence of religious consciousness. **Anthony Freeman**'s article brings together three historically disparate ideas to conceive a naturalistic understanding of 'God as an Emergent Property'. These are: first, the fifth-century Christian notion that the divine/human relation in Christ paralleled the soul/body relation in all humans; secondly, the nineteenth-century view of Friedrich Schleiermacher that the divine presence in Christ was a function of an emergent 'God-consciousness' in his human awareness; and finally a moderate emergentist approach to human consciousness, along the lines of John Searle.

Finally there is a short appreciation and critique by **Alwyn Scott** of Huston Smith's recent book *Why Religion Matters*. Scott draws attention to the contrast between the top-down hierarchy said by Smith to be typical of religion: 'Spirit > soul > psyche > body' and the bottom-up scientific order: 'material > body > psyche > soul'. Smith decries the exclusion of religion by reductive 'scientism', but the absence (pointed out by Scott) of the word 'emergence' from his own index suggests that care is needed on both sides to avoid caricature. But with greater openness, the joint enterprise between science and religion hinted at in the title of Smith's last chapter and adopted as the title of Scott's article ('We Could Be Siblings Yet') might come to pass.

Robert Van Gulick

Reduction, Emergence and Other Recent Options on the Mind/Body Problem
A Philosophic Overview

I: Introduction

Though most contemporary philosophers and scientists accept a physicalist view of mind, the recent surge of interest in the problem of consciousness has put the mind/body problem back into play. The physicalists' lack of success in dispelling the air of residual mystery that surrounds the question of how consciousness might be physically explained has led to a proliferation of options. Some offer alternative formulations of physicalism, but others forgo physicalism in favour of views that are more dualistic or that bring in mentalistic features at the ground-floor level of reality as in pan-proto-psychism.

The situation might be viewed as a case of what the philosopher of science Thomas Kuhn (1962) called *extraordinary science*, i.e., a period of ferment and theoretical experimentation in which the practices and concepts of so-called 'normal science' are loosened in ever more radical ways in response to a persistent anomaly, a problem that resists solution within the limits of the prevailing *normal science* paradigm. Some such episodes usher in full-blown scientific revolutions that sweep away and supplant the prior paradigm; others get resolved more conservatively as ways are found to resolve the anomaly without wholesale abandonment of prior commitments. In those Kuhnian terms, physicalism (particularly the sort of functionalistic nonreductive physicalism that has become the mainstream view among philosophers in recent decades) plays the role of normal science, and consciousness (especially the so-called 'hard problem' of explaining how phenomenal consciousness might be just a physical aspect of reality) provides the anomaly that generates the push toward extraordinary theorizing (Chalmers, 1996). How the current psycho-physical crisis will be resolved as yet remains unclear; revolutions may or may not be needed.

My aim here is to give an overview of the recent philosophic discussion to serve as a map in locating issues and options. I will not offer a comprehensive survey of the debate or mark every important variant to be found in the recent literature. I will mark the principal features of the philosophic landscape that one might use as general orientation points in navigating the terrain.

I will focus in particular on three central and interrelated ideas: those of emergence, reduction, and nonreductive physicalism. The third of these, which has emerged as more or less the majority view among current philosophers of mind, combines a pluralist view about the diversity of what needs to be explained by science with an underlying metaphysical commitment to the physical as the ultimate basis of all that is real. The view has been challenged from both left and right, on one side from dualists (Chalmers, 1996) and on the other from hard core reductive materialists (Kim, 1989). Despite their differences, those critics agree in finding nonreductive physicalism an unacceptable and perhaps even incoherent position. They agree as well in treating reducibility as the essential criterion for physicality; they differ only about whether the criterion can be met. Reductive physicalists argue that it can, and dualists deny it.

The terms 'reduction', 'nonreductive' and 'emergence' get used in a bewildering variety of ways in the mind–body literature, none of which is uniquely privileged or standard. Thus clarity about one's intended meaning is crucial to avoid confusion and merely verbal disagreements. Thus, much of my mapping will be devoted to sorting out the main versions of reduction and emergence before turning to assess their interrelations and plausibility. My intent is to act largely as a guide and not an advocate. Though I am sure my biases will sometimes affect how I describe the issues, my goal is to lay out the logical geography in a more-or-less neutral way.

II: The Varieties of Reduction

The basic idea of reduction is conveyed by the 'nothing more than . . .' slogan. If Xs reduce to Ys, then we would seem to be justified in saying or believing things such as 'Xs are nothing other (or more) than Ys', 'Xs are just special sorts, combinations or complexes of Ys', or 'Xs are nothing over and above Ys'. However, once one moves beyond slogans, the notion of reduction is ambiguous along two principal dimensions: the types of items that are reductively linked and the nature of the link involved. Thus, to define a specific notion of reduction, we need to answer two questions:

- Question of the relata: Reduction is a relation, but *what types of things* does it link?

- Question of the link: *In what way(s)* must the items be linked to count as a reduction?

Let us first consider the question of the relata. Between *what types of things* might the reduction relation hold? The notion gets interpreted in two distinct ways that involve very different sorts of relata. It can be viewed either as

- a relation between real-world items — objects, events, or properties — which we might term *Ontological Reduction* (ONT-Reduction).

or as

- a relation between representational items — theories, concepts or models — which we can call *Representational Reduction* (REP-Reduction).

There are obviously important connections between the two families. But they involve distinct types of relata, and one must not conflate them as too often happens. Speakers in the reduction debate often talk past one another by failing to distinguish ontological from representational notions, especially in interdisciplinary settings that combine scientists and philosophers.

The distinction is crucial as well for locating nonreductive physicalism in the logical space of options. It typically combines a denial of some form(s) of representational reduction with the acceptance of some type(s) of ontological reduction supposedly adequate to secure its physicalist credentials. It claims to coherently conjoin representational nonreduction with an ontological link robust enough to meet the demands of physicalism. It's critics deny that can be done, but the claim at least locates the view in logical space.

The diagram in figure 1 shows the first step in our taxonomy by subdividing the types of reduction into two families based on their answer to our first diagnostic question.

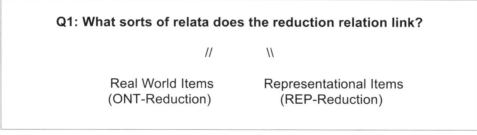

Figure 1

Each family further subdivides based on the specific types of relata involved. Thus, ontological reductions might involve relations between things of various kinds:

- objects in two domains (e.g., minds and brains, or pains and neuron firings),
- properties (e.g., feeling pain and having neural activation of type N_p, or wanting a cup of coffee and being in a neurofunctional state of type Nf_d),
- events (e.g., Bill's having a red visual experience and Bill's brain being globally active in a way that includes neural activities of type N_{rve} in his visual cortex as part of its global focus),
- processes (e.g., my recalling of the cellist's performance and a sequence of reciprocal neural interactions RNI_c between multiple limbic and cortical areas).

REP-reduction similarly divides into more specific subtypes based on the particular relata involved, which might include any of the following:

- concepts (e.g., links between our first-person concept of phenomenal red and concepts from neuroscience),
- theories (e.g., links between theories of conscious experience and theories of global brain function),
- models (e.g., links between models of consciousness and models of reciprocal brain activity),
- representational frameworks (e.g., links between the phenomenal first-person descriptive/explanatory framework and third-person neuroscience frameworks).

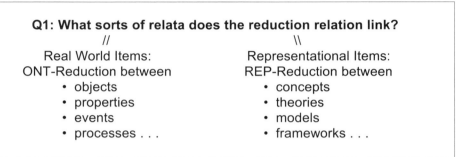

Figure 2

Let us move on to our second diagnostic probe, the question of the link: *In what way(s)* must the items be *linked* to count as a case of reduction? Again, there are a variety of answers on both the ontological and the representational side. With regard to ONT-reduction, the question becomes:

Question of the *ontological link:* How must things be related for one to ontologically reduce to the other?

At least five major answers have been championed in the literature:

- elimination
- identity
- composition
- supervenience
- realization

The relative merits and faults of the competing proposals have been extensively and intricately debated, but for present purposes it should suffice to say a brief bit about each and give a general sense of the range of options.

Elimination. One of the three forms of reduction listed by Kemeny and Oppenheim in their classic paper on reduction (1956) was replacement, i.e., cases in which we come to recognize that what we thought were Xs are really just Ys. For example, we've come to see that what had been thought of as demon possession is just a form of psychosis, perhaps a type of schizophrenia with auditory hallucinations. In such a case we might say that demon possession has been reduced to a mental illness; schizophrenia replaces demon possession in our inventory of the world. Eliminativists believe that a similar fate awaits many of the commonly alleged denizens of the mental domain; qualia, beliefs, intentionality and even consciousness have all been slated for eventual replacement by supposedly more mature scientific alternatives (Rorty, 1970; Churchland, 1981; Dennett, 1988; Wilkes, 1988; 1995). It is difficult to prove or refute predictions about the future course of inquiry and how it will affect our future conceptual repertoire and beliefs about what is and isn't real. Nonetheless, the eliminativist shoulders a nontrivial burden to motivate his claim that so much of what we take to be real in the mental domain might go the way of demons or black bile. Our sense of what is mentally real seems too intimate and too useful to turn out to be so badly mistaken as to justify an eliminativist judgment. The debate in the literature is extensive, but in this context it is enough just to note the controversial status of the issue.

Some contemporary readers may find it odd to describe elimination as a form of reduction since according to the eliminativist there is nothing really there to be reduced. If there are no beliefs, how can they be reduced to anything? Nonetheless, I believe it is appropriate to keep it on our list. As noted above, reduction by replacement, which is what elimination basically involves, was listed early on by Kemeny and Oppenheim in their seminal work (1956) as one of the main types of reduction. Moreover, the eliminativist view is one way to unpack the basic reductive notion of 'nothing but . . .'; demon possession (or what we thought of as demon possession) turns out to be nothing but organic brain disorder.

Identity. Identity falls at the opposite extreme from elimination. It involves cases in which we continue to accept the existence of Xs but come to see that they are identical with Ys (or with special sorts of Ys). Xs reduce to Ys in the strictest sense of being identical with Ys. This most often happens when a later Y-theory reveals the true nature of Xs to us. We have come to see that heat is just kinetic molecular energy, that lightning is just an atmospheric discharge of static electricity and that genes are just functionally active DNA sequences. However, the 'just are . . .' locution in such cases does not lead us to eliminate or deny the existence of the prior items; we do not deny the reality of lightning, heat or genes. Rather, we see that two distinct reference routes converge on the same item. In the logician Frege's famous example of the evening star and the morning star, both turned out to be the planet Venus. (Frege, 1892) So, too, identity theorists claim mental states, events and properties will turn out to be identical with neuroscientifically discovered items. Those mental states are every bit as real as heat and genes, but their true nature is yet to be discovered by a more mature mind/brain theory.

Contemporary physicalism first developed as an identity theory in the 1950s and 60s (Place, 1956; Smart, 1959), specifically as what has come to be called the type–type central state identity theory, which held that types of mental states (e.g., being a stabbing pain) were identical with types of states within the central nervous system (e.g., c-fibre firings or, more realistically, certain patterns of firing in anterior cingulate, somato-sensory cortex and interconnected limbic areas). The theory fell out of favour quickly for a variety of reasons, such as the multiple realizability objection which appeals to the fact that one and the same type of mental state might be realized by neurally quite dissimilar structures in different creatures, in different humans, or even in the same person at different times (Putnam, 1972; Kim, 1993b). Thus, by the late 1960s, most physicalists had moved away from the type–type identity theory in favour of some form of functionalism that treats mental states and properties as higher-order features defined by their higher-level roles but nonetheless realized solely by their neural substrates, much as the higher-level program states of a computer are realized by the underlying states of its hardware.

The situation is actually a bit more complicated since some functionalists (Lewis, Armstrong, . . .) identify the mental state-type not with the higher-order or role property but with the specific structural property that plays that role in a given species or population. Though they pick out the property by the role it typically plays, they set the identity conditions in structural terms. In that respect they are more like classic type–type identity theorists than like other functionalists who set the identity conditions in terms of the role itself rather than in terms of the typical occupant of the role. According to the occupant functionalist (e.g., Lewis), if neural state N typically plays the role in humans associated with having a desire for coffee, and I am in state N, then I have a desire for coffee even if in my particular case N exhibits none of the causal roles associated with such a desire, e.g., it does not make me more likely to accept or drink a cup of coffee if offered one or to express my desire for coffee if asked if I'd like some. Many functionalists find such claims counterintuitive; Lewis concedes that such cases would indeed be odd but nonetheless argues they would be still be cases of my having coffee desires despite their nonstandard causal profiles. We need not decide the issue here; it is enough to note that differing views can be found in the literature.

The type–type identity theory has enjoyed a recent though modest rebirth of interest (Hill and McLaughlin, 1999). Some philosophers have looked to it as a means of solving or dissolving the supposed explanatory gap that confronts and baffles those who try to explain how and why any given conscious state correlates with or might be realized by a given neural state. These neo-identity theorists argue that because it is an identity that is involved there is no explanatory gap to bridge. There are not two things whose linkage needs to be explained; there is just one thing, and it like everything else is necessarily identical with itself and not with anything else. If Brian's pain just *is* a certain pattern of brain activity in the *identity* sense of 'is', then there is no gap to be closed any more than there is any case of identity. If there is just one thing, then as a matter of simple logic it's the same as itself. Some have complained that the explanatory lacuna merely reappears as the unsatisfied demand for some account of how the mental and physical pathways might converge on a common referent, and that complaint seems justified. Nonetheless, it's important to acknowledge that identity versions of reductive physicalism, though far less popular than a few decades back, are still alive and being actively defended.

Composition. One seemingly plausible alternative to identity is composition. If mental things (e.g., minds) are *composed* entirely of physical parts, might that not suffice to justify the reductionist claim that they reduce to the physical or are all 'just physical'. If all of a thing's parts are physical, can the reductionist not say that it 'contains nothing over and above the physical'? Moreover, composition is distinct from identity and has a different logical status that easily accommodates the sort of multiple realizability objections raised against identity theory. Higher-level objects can outlive the components of which they are at a given time composed. A marching band which gradually changes its membership over time can continue to exist even after one hundred per cent of its original members have retired; identity of the band itself is not dependent on the sameness of its underlying composition. So, too, it might seem a given mental state might persist through changes in its underlying neural composition. There may be good mentalistic reasons for regarding the memory that I have on Tuesday of the film I saw on Sunday as the same mental state as the memory that I had of it on Monday, even if the two differ nontrivially in their neural components. Brains and the patterned information they encode seem to be quite neurally dynamic; retaining or recalling a given memory need not rely upon the very same ensemble of neuron activations.

Despite its obvious attractions, composition will not suffice to ground reductive physicalism. Indeed it is compatible with various forms of property dualism, including both fundamental and emergent property dualism. To say that a thing is composed *entirely of physical parts* is not the same as saying that *all its parts are entirely physical*; to assert the former is to say only that all its parts have physical properties, but the second asserts as well that those parts have *only* physical properties. Without the second stronger assertion, nothing excludes the possibility of nonphysical mental properties either at the level of the parts or at the level of the whole (emergent property dualism), and the appeal to composition *per se* will not suffice to support that stronger assertion.

Supervenience. Some philosophers in recent years have appealed to the notion of supervenience as a way of getting beyond mere composition and reductively linking mental and physical properties themselves while still stopping short of strict identity (Davidson, 1970; Kim, 1982). The issue is complicated by the lack of any philosophical consensus about the identity and individuation conditions for properties; there is no dominant view in the field about the metaphysical status of properties and the conditions under which one property is the same as another. Nonetheless, various attempts have been made to analyse the relation between mental and physical properties in ways that might legitimate physicalism without invoking a strict type–type identity of properties.

Supervenience, which involves the dependence of one set of properties on another, was first proposed early in the twentieth century as a way of explaining the relation between normative properties such as moral and aesthetic properties and their non-normative bases (Moore, 1902). The basic idea is that one set of properties (X-properties) supervenes on another (Y-properties) such that there can be no X-differences without Y-differences, or to put the point the other way round, any two things sharing all their Y-properties must also be alike in all X-properties. For example, the beauty of a painting may not be identical with any of its strictly physical properties such as its distribution of pigments on the surface of the canvas, but any other painting that shared all the physical properties of the first would also have to share its aesthetic properties. If the first were sublimely beautiful, so too would be the second. In the mind/body domain, the basic view can be conveyed by the slogan, 'No mental difference without a physical difference' (Davidson, 1970; Kim, 1982). Although supervenience theorists have not generally labelled themselves as reductionists, the dependence relation provides one way to cash out the basic reductionist idea that 'Xs are just Ys'.

Although supervenience enjoyed a brief period of intense interest as a possible way of making sense of ontological physicalism, it has now generally fallen out of favour. Even Jaegwon Kim, who played the largest role in bringing the notion to the centre of discussion (1982; 1993a) has acknowledged more recently (1999) that supervenience is too weak a relation to validate physicalism and is *a fortiori* an inadequate way of analysing the concept of reduction. Kim, for example, has conceded that supervenience is compatible with both property dualism and dual aspect theory. Even if the mental and physical realms were distinct and separate, supervenience could still hold as long as there were invariant correlations between the two — whether underwritten by natural law (*nomic supervenience*) or some stronger metaphysical link (*metaphysical supervenience*). Indeed, the first sort of correlation would hold in the world of a classical interactive dualist like Descartes and the latter in the world of a panpsychic monist like Spinoza. If supervenience is compatible with such explicitly nonphysicalist views, it seems unlikely to provide an adequate account of physicalism and even less of how the mental might reduce to the physical. Some (e.g., Chalmers, 1996) have replied that what is needed is a relation of *logical supervenience*, according to which as a matter of logic it is impossible that mental and physical properties might independently vary. But then the question quickly arises of what sort of link short of identity might underwrite such a logically necessary link. Thus, if an adequate answer is to be found, it looks like the real explanatory work will have to be done by the story of that underlying relation (whatever it might be) rather than by the notion of supervenience *per se*.

Realization. One way of spelling out that underlying story might be in terms of realization. The multiple realizability of mental states was used as an objection against type–type identity theory (Putnam, 1972), but it has also provided the basis for a positive view of the psycho–physical link that might suffice as a form of ontological reduction. Realization is especially attractive to functionalists. Two systems can manifest the same functional property even if it is realized by different structures in the two cases. This holds both for natural biological functions and artefactual ones. Nature can typically build a membrane with a given permeability profile in more than one way, and engineers can design a signal amplifier with the same input/output function using many different hardware setups. However, as an ontological matter, the given functional property is fully realized in each case by its underlying physical components and their mode of composition. Once you've fixed all the facts about the structures and processes at the physical level, the facts about the functional properties follow automatically.

Realization appeals to those who favour a mind/computer analogy (Putnam, 1972; Fodor, 1981) since one and the same software or computational processes can be realized by many different types of hardware. However, realization is also invoked by some philosophers like John Searle (1992), who are anti-computationalist and explicitly nonfunctionalist about the mental. Searle denies that mental properties can be functionally analysed because of what he regards as their irreducible first-person intrinsic features, but he nonetheless classifies his view as physicalist by appeal to his slogan that mental properties are 'caused by and realized in' the physical processes of the brain. For Searle, the realization claim is essential to avoid property dualism, which would remain an option if he claimed merely that mental properties were caused by physical brain processes. But with the addition of the realization requirement, he regards himself as having shown mental states as metaphysically no more problematic than the liquidity of room-temperature water, which is analogously *caused by and realized in* interactive collections of H_2O molecules.

Realization plays a role as well in many versions of nonreductive physicalism. In her attempt to combine ontological physicalism with a denial of representational reduction (ONT-Reduction & Not REP-Reduction), the nonreductivist most often appeals to realization to secure her credentials as an ontological physicalist (Van Gulick, 1992). Can she succeed? Can she give an account of a realization that is strong enough to vindicate physicalism, yet consistent with a robust denial of REP-reduction? We will get to that question below, but first we need to shift our attention from the ontological side (ONT-Reduction) and consider the equally diverse family of relations that qualify on the representational side as kinds of REP-reduction.

In a case of REP-reduction, one set of representational items is reduced to another. In answer to Question 1, we noted above that REP-reductive relations might hold among at least four different kinds of representational items: concepts, theories, models and frameworks. Those four do not exhaust the relevant options, but for present purposes we can restrict ourselves to them. What then of the representational version of Question 2:

Question of the *representational link:* How must things be related for one to representationally reduce to the other?

There are a diversity of answers given in the literature, some of which make sense only with respect to certain kinds of items in the representational domain; some relations, such as derivability, might make sense as relations between theories but not between models or among concepts. However, certain commonalities run through the family of REP-reductive relations. They all involve some sort of *intentional equivalence*, i.e., some correspondence in terms of what they can or do represent as opposed to a mere correspondence in their form or intrinsic properties. The basic idea is that one representational item (or set of items) reduces to another just if the first is linked to the second in terms of what it can or does say about the world and its features. Thus, REP-Reductive relations generally concern either the comparative expressive powers of representational items (what they *can say*) or correlations in their assertoric content (what they *do say*). To invoke a hoary and controversial traditional division, representational reductions turn on either relations of meaning or relations of truth. Most of the specific variants of REP-Reduction fall into one of five general categories:

- Replacement,
- Theoretical–Derivational (Logical Empiricist),
- *A priori* Conceptual Necessitation,
- Expressive Equivalence (two-term semantic relation),
- Teleo–Pragmatic Equivalence (n-term pragmatic relation).

These alternatives have been extensively debated in the literature, but I will say only enough about each to indicate the range of options that are active in the field.

Replacement. The analogue on the representational side of elimination on the ontological side would be replacement. Our prior ways of describing and conceptualizing the world might drop out of use and be superseded by newer, more adequate ways of representing reality. For example, many of our mentalistic concepts might turn out not to do a good job of characterizing the aspects of the world at which they were directed, as has happened with demon concepts. If so, future science might develop alternative concepts and theoretical resources that would more accurately and effectively represent reality. If adopting those newer systems of representing should lead us to drop our former mentalistic outlook, then in an extreme sense our mentalistic way of speaking and thinking might be regarded as having been reduced to its representationally superior replacement. However, most notions of REP-Reduction to be found in the literature are more conservative and involve preserving more of the truth or expressive content of the reduced theory.

Theoretical–Derivational. The classic notion of reduction in terms of theoretical derivation, as found in Kemeny and Oppenheim (1956) or in Ernest Nagel's classic treatment (1961), descends from the logical empiricist view of theories as interpreted formal calculi statable within the resources of formal symbolic logic. Given the axioms or laws of such a theory together with a formally statable set of actual conditions, one can derive all its consequences, observational and otherwise, by working out its formal implications. Thus, if one such theory T1 could be logically derived from another, T2, then everything T1 says about the world would be captured by T2, and T1 could be said to reduce (REP-reduce) to T2. Because the theory to be reduced, T1, normally contains terms and predicates that do not occur in the reducing theory T2, the derivation also requires some bridge laws or bridge principles to connect the vocabularies of the two theories. These may take the form of strict biconditionals linking terms in the two theories, and when they do, such biconditionals may underwrite an ontological identity claim. If a gas has a given temperature when and only when its molecules have a given average kinetic energy, then we may go on to infer that temperature just *is* average kinetic energy, and that heat is *identical with* molecular motion. However, the relevant bridge principles can also take other forms; they need not be strict biconditionals. All that is required is enough of a link between the vocabularies of the two theories to support the necessary derivation.

One other caveat is in order. In most cases what is derived is not strictly speaking the original reduced theory but an image of that theory within the reducing theory, and that image is typically only a close approximation of the original rather than a precise analogue (Churchland, 1985). For example, in the paradigm case of the reduction of classical thermodynamics to statistical mechanics, the image of the classical laws within the statistical domain allows for possible though extremely improbable deviations from the classical laws. Thus, as a strict matter of logic, those prior laws are not derived. However, we accept that the classical laws were not strictly speaking true, and the match between the original laws and their statistical analogues is so close that we accept the former as having been reduced to the latter.

Were we to try to apply the theoretical–derivational model of REP-reduction to the mind/body case, we would need to find a set of bridge principles that allowed us to derive all the truths of our mentalistic theories of consciousness from the laws and statements of the relevant reducing theory (theories) whatever they might be: neurophysiological, computational, quantum mechanical or otherwise. As in the thermodynamics case, the derived result might not be an exact analogue, but it would have to be a close enough approximation to our original pre-reductive mentalistic theory to justify the claim of REP-reduction by theoretical derivation.

Such a prospect might have once been viewed as the likely result of eventual scientific progress; logical empiricist views of the unity of science envisioned such an eventual formal integration of our scientific representation of reality (Oppenheim and Putnam, 1958). At present, far fewer philosophers expect such derivations to be produced even in the long term. Ontological dualists believe that our mental and physical theories describe different domains, and thus they expect no reduction of the theories that describe them. But REP-reduction is deemed unlikely also by many physicalists, especially those who accept some form of nonreductive physicalism. Unlike the dualists, they believe there is just one domain of reality which at some level is correctly represented by physics, but they also believe that adequately representing all the complex features that that reality exhibits at its many different levels requires the use of a diversity of theoretical and representational resources beyond those provided by the formal structures of physical science *per se* (Fodor, 1974; Boyd, 1980; Garfinkel, 1981; Van Gulick, 1992). We will inquire further below about why nonreductive physicalists believe that, and about whether such a view can be consistently combined with ontological physicalism. For the present we need only note that for a diversity of reasons, theoretical REP-reduction by derivation is rejected in the mind/body case by many but not all current philosophers.

A Priori Conceptual Necessitation. Though many reject theoretical reduction as too strong a link, other philosophers regard it as too weak for adequately REP-reducing mind to body. In particular, they demand that the bridge principles not merely support the derivation of the reduced theory but do so by linking the two theories via necessary and *a priori* conceptual links or inter-theoretical definitions (Levine, 1983; 1993; Chalmers, 1996). They would not count a derivation based upon mere empirically established links or biconditionals as a successful reduction, since such empirical bridge principles might only describe the correlations between properties in distinct and separate mental and physical domains. This alleged weakness of the derivational account of reduction is the representational analogue of the ontological faults charged against supervenience. Lawlike covariance, though sufficient to support nomic supervenience on the ontological side and inter-theoretical derivation on the representational, nonetheless seems inadequate to vindicate physicalism or reductionism because it is compatible with both property dualism and dual aspect theory as long as the mental and physical domains are lawfully linked; something which even a Cartesian dualist accepts. Proponents of the *a priori* view argue that nothing short of logically sufficient conceptual connections will suffice for one theory to reduce another (Chalmers, 1996). They often couple that claim with a demand for reductive explanation, i.e., for an explanatory account that lets us see in a conceptually necessary way how the conditions described by the reducing theory must as a matter of logic alone guarantee the satisfaction of those described by the reduced theory. They appeal to models such as that of the liquidity of water. The explanation in such cases supposedly includes two components:

- first, an analysis of the concept to be reduced (e.g., liquidity) in terms of some set of typically functional conditions, and
- second, an account of how those conditions would as a matter of mere logic be satisfied by any underlying system meeting the conditions described by the reducing theory (the micro-interactions of the collection of room-temperature H_2O molecules).

The 'apriorists' contend that nothing less would suffice as a theoretical reduction of the mental to the physical.

Once again, the issues are numerous and debates in the literature extensive, but we need only note the controversial nature of the apriorists' claim. Though the *a priori* view has intuitive appeal and its share of supporters, it also has a host of critics (Block and Stalnaker, 1999; Van Gulick, 1999; Yablo, 1999), some of whom plausibly charge it with setting up — or at least implying — a false dilemma. Most physicalists would agree that mere brute fact mental–physical correlations unsupported by appeal to any underlying explanation of why they were so linked would fall short of providing a reduction even if they sufficed as a bridge to derive our mental theories from our physical ones. Indeed, right at the start of contemporary physicalism, the philosopher Herbert Fiegl (1958) disparaged such unsupported links as 'nomological danglers'. However, demanding that the bridge principles provide logically necessary and *a priori* conceptual links between the two domains seems to swing to the opposite extreme. Surely there must be intermediate cases that involve explanatory rather than merely brute links, but that nonetheless fall short of the apriorists' radical requirement for strict logical entailments. There is no consensus about what would count as an adequate explanation of the psycho–physical link and about what sorts of bridge principles would suffice for a satisfactory theoretical derivation, but many physicalist believe that the answer lies somewhere between the two extremes (Van Gulick, 1992; Kim, 1999).

Expressive Equivalence. The two versions of REP-Reduction considered thus far both concern what we might call truth preservation, i.e., in both cases everything the reduced (mental) theory says about the world is also asserted by the reducing (physical) theory combined with the requisite sorts of bridge principles. However, REP-Reduction might be viewed not as a matter of equivalence in what *is said* about the world, but as merely a matter of preserving the expressive range of what one *can say*. One representational system, R1, might be regarded as reducible to another, R2, as long as every state of affairs representable by R1 could also be represented by R2. The expressive range of R1 would be entirely contained within that of the reducing system R2.

Though expressive equivalence seems to set a weaker requirement than the derivational or apriorist versions of REP-Reduction, it is still far from obvious that such a criterion can be met. Many philosophers, dualist and otherwise, have appealed in particular to the alleged special nature of first-person phenomenal concepts and the sorts of experiential facts that we can supposedly know or understand only through their use. Drawing on an old empiricist intuition that goes back at least to John Locke (1688) in the seventeenth century, contemporary philosophers such as Thomas Nagel (1974) and Frank Jackson (1982; 1986) have argued that there are facts about consciousness that can be adequately known or understood only from the first-person experiential perspective; for example, one can fully understand what it's like to taste a pineapple only if one has oneself had such an experience. (Although Jackson himself has recently [1998] changed his mind on this issue, many others continue to invoke his original position and have not followed him in his reversal of opinion.) Thus, some philosophers claim that physical theory with its reliance on third-person concepts can never fully achieve the expressive range of our mental modes of representing, especially those that involve experiential concepts. If there are indeed subjective facts that lie beyond the representational power of physical theory, then it may be impossible to REP-Reduce the mental to the physical even in the weaker sense of expressive equivalence. There may be things that we can say mentalistically about the world that fall outside the range of what can be said using the resources of physical theory. Unsurprisingly, philosophers disagree about whether or not there are such facts (Lewis, 1982; Churchland, 1985; Van Gulick, 1985; 1993a; Levin, 1986; Loar, 1990; Lycan, 1990), but for present purposes we need only note that if there are, then that would seem to preclude REP-Reducing the mental to the physical in the expressive equivalence sense.

Teleo-Pragmatic Equivalence. The expressive equivalence account of REP-Reduction treats representation as primarily a two-term relation between a representation (or set of representations), R, such as words or sentences in a language or theory, and the item (or items), X, represented by R. Moreover, the relation is thought of in terms of familiar semantic notions such as reference and meaning. Thus, if we interpret REP-Reduction as a matter of expressive equivalence, the question becomes: For every representing element of T1 that means M or refers to an item X, is there a corresponding element or combination of representing elements in T2, call it R*, that has that same meaning or referent? This is certainly one legitimate way to define REP-Reduction, but it may not be the most helpful or revealing. Though representation can be viewed as a two-term relation, doing so involves abstracting away from other significant parameters of representation that are likely to be of importance to understanding the REP-reductive or non-REP-reductive nature of the mind/body link. In particular, what a given representation, R, succeeds in representing is crucially dependent upon the causal structure of the representation-user, U, the social and physical context, C, in which R is applied, and the modes of causal and epistemic access that R affords to U when used in C. Thus, rather than being just a two-term relation between a representation, R, and represented item, X, representation is at least a four-term relation that includes as well a representation user, U, and a context of use, C; nor need we stop at four, additional parameters might easily be added such as the goals or ends toward which the representation is to be applied. Reframing the question of REP-Reduction in terms of this more complex relation, the question becomes as follows: If U can use R from theory T1 in context C to represent X, is there an R* from T2 that U can similarly use in C to represent X, or at least some R† from T2 that U can use in some context C† to represent X? To put the matter less abstractly, the problem is to find a way, if possible, for us to use the contextually embedded resources of the reducing theory to do the equally contextual representational work done by the items in the theory we are trying to reduce. Nor should we ignore the third or pragmatic parameter mentioned above. Success in real-world representation is in large part a practical matter of whether and how fully our attempted representation provides us with practical causal and epistemic access to our intended representational target. A good theory or model succeeds as a representation if it affords us reliable avenues for predicting, manipulating, and causally interacting with the items it aims to represent. This is no less true in the natural realm when one is judging that a given structure in the rat hippocampus serves as allocentric representation of its spatial environment than it is when we judge that an economist's model has succeeded in representing the effect of interest rate changes on housing markets. In both cases, it is the practical access that the model affords to its user in its context of application that justifies us in viewing it as having the representational content that it does (Van Gulick, 1992).

Viewed from this more inclusive contextual and pragmatic perspective, the question of REP-Reduction becomes one about the ability of representation-users to gain the same modes of access with the alleged reducing representations that they do with the representations to be reduced. In the particular mind/body case, this turns in part on our ability (or inability) to use the representational resources of physical theory to replicate the functional interactive profile of the access afforded us by our mentalistic resources including our first-person concepts and theories. Put in this way, REP-Reduction may seem far less plausible; it seems unlikely that we could use physical theory to gain access to our mental states and processes in ways that afford us the same understanding that we achieve through our first-person and introspective modes of representation. The differences in the contexts of use are so great that they alone seem sufficient to make such an equivalence for all practical purposes impossible. Many of our first-person concepts are so directly embedded within our intra-mental processes of self-regulation, self-monitoring and self-modulation that it is difficult to see how any third-person system of concepts provided by physical theory could achieve a pragmatic profile sufficiently similar to support a claim of REP-Reduction. Indeed, it is for just that reason that philosophers who adopt a pragmatic view of representation typically also deny the possibility of REP-Reduction and are in that sense nonreductionists (Putnam, 1978; Garfinkel, 1982; Van Gulick, 1992).

Though much more could be said about the many varieties of ontological and representation reduction and their respective faults and merits, I hope at least to have surveyed the main versions of each as graphically summarized in figure 3.

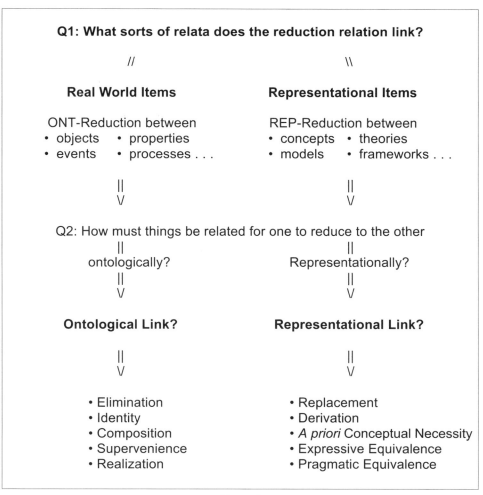

Figure 3

III: The Varieties of Emergence

We must thus turn our attention to the notion of emergence, which like reduction gets interpreted in diverse ways in the mind/body literature (Searle, 1992; Hasker, 1999; Silberstein and McGeever, 1999). Again, my aim will be to survey the main variants. Only when that has been done can we investigate the relations among the many members of these two diverse families of concepts.

The basic idea of emergence is more or less the converse of that associated with reduction. If the core idea of reduction is that Xs are 'nothing more than Ys' or 'just special sorts of Ys', then the core idea of emergence is that 'Xs are more than just Ys' and that 'Xs are something over and above Ys'. Though the emergent features of a whole or complex are not completely independent of those of its parts since they 'emerge from' those parts, the notion of emergence nonetheless implies that in some significant and novel way they *go beyond* the features of those parts. As we will see, there are many senses in which system's features might be said to emerge, some of which are quite modest and unproblematic (Searle, 1992) and others which are more radical and controversial (Hasker, 1999).

The varieties of emergence can be divided into several groups along lines that are similar in at least some respects to the divisions among the types of reduction. For example, emergence relations might be viewed either as objective metaphysical relations holding among real-world items such as properties, or they might be construed as partly epistemic relations that appeal in part to what we as cognitive agents can explain or understand about such links.

```
Q1: What sorts of factors figure in emergence relations?

           //                              \\

Metaphysical: relations         Epistemic: cognitive explanatory
among real world items          relations about real world items
```

Figure 4

Relations of the first sort are objective in the sense that they concern the links or lack of links between real items such as properties, independent of any considerations about what we humans, as epistemic subjects, can or cannot explain or understand about them. Relations of the second sort are epistemic, and in a sense subjective, because they turn crucially on our abilities or inabilities to comprehend or explicate the nature of the links or dependencies among real-world items rather than just on those links alone. The two sorts of notions often get run together in the discussion of emergence, but it is important to keep them distinct just as with the ontological and representational notions of reduction discussed above (Silberstein and McGeever, 1999).

REDUCTION & EMERGENCE: A PHILOSOPHIC OVERVIEW

On the objective side, two main classes of emergents can be distinguished: properties and causal powers or forces. The distinction between the two is not sharp and involves possible overlaps, especially if one individuates properties in terms of their causal profiles. Nonetheless, the issue of emergent causation is critical, so it's worth distinguishing at least initially between emergent properties and emergent powers, even if the line subsequently blurs a bit. Within each of the two classes, there is a continuum of cases running from the extremely modest to the extremely radical. The former involve emergent features, whether properties or powers, that are very similar in nature to the features from which they emerge, whereas the emergents in the latter sorts of cases are most unlike their nonemergent bases. Although the cases differ by many variations of degree, we can give some sense of their range by dividing the cases into three rough categories of increasing radicality which we can label: specific-value emergence, modest-kind emergence, and radical-kind emergence.

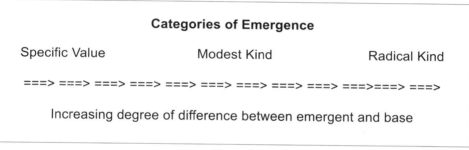

Figure 5

We can define the three roughly as follows.

- Specific Value Emergence. The whole and its parts have features of the *same kind*, but have different *specific subtypes or values* of that kind. For example, a bronze statue has a given mass as does each of the molecular parts of which it is composed, but the mass of the whole is different in value from that of any of its proper material parts.

- Modest Kind Emergence. The whole has features that are *different in kind* from those of its parts (or alternatively that *could* be had by its parts). For example, a piece of cloth might be purple in hue even though none of the molecules that make up its surface could be said to be purple. Or a mouse might be alive even if none of its parts (or at least none of its subcellular parts) were alive.

- Radical Kind Emergence. The whole has features that are both
  1. different in kind from those had by its parts, and
  2. of a kind whose nature and existence is not necessitated by the features of its parts, their mode of combination and the law-like regularities governing the features of its parts.

Whether or not there are any cases of radical-kind emergence is controversial. Physicalists who would readily concede the reality of specific-value emergence and modest-kind emergence would most likely deny there are cases of the more extreme sort. Accepting radical-kind emergence would be conceding that there are real features of the world that exist at the system or composite level that are not determined by the law-like regularities that govern the interactions of the parts of such systems and their features. Doing so would require abandoning the atomistic conception, which is typically embraced by mainstream physicalism. Radical-kind emergence would require giving up at least one of two core principles of atomistic physicalism (AP):

- AP1. The features of macro items are determined by the features of their micro parts plus their mode of combination. (In a slogan: Micro features determine macro features.)
- AP2. The only law-like regularities needed for the determination of macro features by micro features are those that govern the interactions of those micro features in all contexts, systemic or otherwise.

The idea of AP2 is that there are no laws governing micro features in systemic contexts other than those that govern them outside such contexts. The intent is to exclude special laws that come into play only in restricted systemic contexts. Once the micro features and their distribution have been fixed, the micro necessities by themselves suffice to determine the macro properties of the system. No further laws or law-like regularities are needed to necessitate the macro outcomes. Mainstream atomistic physicalism thus includes a commitment to what might be called 'bottom-up determination'. To borrow an example from the philosopher Saul Kripke (1972), once God had done the work of fixing the micro features and laws of the universe, there was no work left to do; in fixing the world's micro nature, He had already determined all its macro properties as well. Or so, at least, atomistic physicalists would claim. As we will see below in discussing the dual revolutions position, some physicalists believe we should give up our commitment to atomism and micro-physical determination in favour of a more holistic view of physical reality; such an alternative though not inconsistent with physicalism *per se* is certainly far from the mainstream view, and it will best to consider it below under the category of other options.

If we cross-pair our three rough divisions along the continuum of unlikeness between emergents and bases with our two classes of objective emergents (properties and powers), we then get six versions of metaphysical emergence as shown in figure 6.

The notion that causal powers might exhibit radical-kind emergence merits special attention since it poses perhaps the greatest threat to physicalism. If wholes or systems could have causal powers that were radically emergent from the powers of their parts in the sense that those system-level powers were not determined by the laws governing the powers of their parts, then that would seem to imply the existence of powers that could override or violate the laws governing the powers of the parts; i.e., genuine cases of what is called 'downward causation'

Q1: What sorts of factors figure in emergence relations?

                //   \\

Metaphysical: relations among real world items   **Epistemic**: cognitive explanatory relations about real world items

‖
V

- **Emergent Properties**
- Specific Value Emergence
- Modest Kind Emergence
- Radical Kind Emergence

- **Emergent causal Powers**
- Specific Value Emergence
- Modest Kind Emergence
- Radical Kind Emergence

Figure 6

(Sperry, 1983; 1991; Kim, 1992; 1999; Hasker, 1999) in which the macro powers of the whole 'reach down' and alter the course of events at the micro level from what they would be if determined entirely by the properties and laws at that lower level. All that is needed to get such a result is the relatively uncontroversial claim that macro-level causal powers have micro effects. That in itself need cause no trouble for the physicalist as long as the macro powers are themselves determined by the micro powers (Kim, 1999). But if some macro causal powers were radically emergent, that would free them of determination by the underlying micro powers thus allowing them to alter the course of micro events in ways independent of the micro-level laws. If the physicalist wants to avoid violation of underlying physical laws, she can allow macro-level properties to have micro effects only if those macro powers are themselves constrained by the micro-level laws, which is of course just what the radical emergence of causal powers denies (Hasker, 1999). It is in this respect that radically emergent causal powers would pose such a direct challenge to physicalism, since they would threaten the view of the physical world as a closed causal system; i.e., the idea that nothing outside the physical causally affects the course of physical events (Kim, 1990; 1999). Unsurprisingly, this feature that makes radical causal emergence so threatening to physicalists (e.g., Kim, 1992; 1999) is the very one that make it so attractive to those emergentists like William Hasker (1999) who invoke emergence in support of ontological dualism.

The challenge to those who wish to combine physicalism with a robustly causal version of emergence (Sperry, 1991; Van Gulick, 1992) is to find a way in which higher-order properties can be causally significant without violating the basic causal laws that operate at lower physical levels. On one hand, if they override the micro-physical laws, they threaten physicalism. On the other hand, if the higher-level laws are merely convenient ways of summarizing complex micro-patterns that arise in special contexts, then whatever practical cognitive value such laws may have, they seem to leave the higher-order properties without any real causal work to do (Kim, 1992). One possible solution would focus on the respect in which higher-order patterns might involve the selective activation of lower-order causal powers. Micro-properties that were causally irrelevant in most configurations, e.g., because their random actions cancelled out and had no significant overall effect, might exert a powerful causal influence in a small range of cases involving higher-level patterns that brought those micro-powers into a coherent mode of action making a major difference to the overall operation of the system (Van Gulick, 1993b). For example, the magnetic fields of molecules can be ignored in explaining most materials since the are randomly aligned. However, when they are coherently oriented in a magnet, those micro-properties are crucial to understanding the causal powers and activity of the system that contains them. Such selectional models need pose no problem for atomistic physicalism since they involve no violation of underlying micro-physical causal laws. However, whether they accord a sufficiently potent form of causal efficacy to macro or system-level properties to justify a claim of emergent causality is open to debate. The selective activation model nonetheless provides at least an example of how one might try to find a way of reconciling physicalism and causal emergence.

Having given at least six rough options on the metaphysical side, let us turn our attention to epistemic notions of emergence. What makes all such notions epistemic is that they involve some respect in which we are unable to predict, explain, or understand the features of wholes or systems by appeal to the features of their parts. Emergence in this sense is thus at least in part subjective, i.e., a matter of our cognitive and explanatory capacities and limits rather than just a matter of relations between objective items as in the metaphysical cases. For present purposes, it will probably suffice to distinguish two versions of epistemic emergence, which focus respectively on different cognitive abilities — the first on prediction and explanation and the second more on representation and understanding. Once again, the lines between the two are not sharp, but worth distinguishing.

Predictive/Explanatory Emergence: Wholes (systems) have features that cannot be explained or predicted from the features of their parts, their mode of combination, and the laws governing their behaviour.

Representational/Cognitive Emergence: Wholes (systems) exhibit features, patterns or regularities that cannot be represented (understood) using the theoretical and representational resources adequate for describing and understanding the features and regularities of their parts.

Both versions of epistemic emergence come in weak or restricted forms and in strong or unrestricted forms. On one hand, the relevant cognitive inability to explain or represent might be a restricted fact about our specifically human limitations, or even more restrictively about our present state of theorizing and scientific progress. Alternatively, the cognitive inability might concern a more general limit that applies universally to all cognitive agents or to all those in some broad category of which we humans are just one example among many. The philosopher Colin McGinn, for example, has argued that humans lack the ability to form the sorts of concepts needed to make the psycho–physical link intelligible (McGinn, 1989; 1991). Thus, he believes it will ever remain a mystery to us, even though he accepts consciousness as an aspect of physical reality and allows that cognizers with concept-forming abilities quite different from our own may be able to comprehend the link in intuitively satisfying ways. It's not that we humans are not smart enough; it is that we have the wrong sorts of minds for solving the psycho–physical puzzle. From McGinn's perspective, consciousness is epistemically emergent in the explanatory sense, at least relative to humans and other cognitive agents with our sorts of conceptual capacities.

We can summarize our quick survey of the varieties of emergence in figure 7.

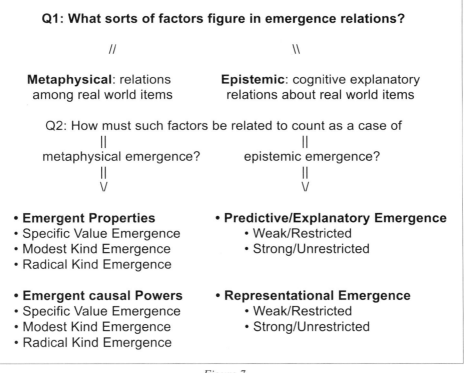

Figure 7

IV: Other Recent Mind–Body Options

Our taxonomizing survey has thus distinguished among ten varieties of emergence and at least ten versions of reduction. Even so, it probably does not capture every interesting variant to be found in the literature, but I hope it includes all the major ones and gives a fair sense of the range of options being discussed in both families. There are, of course, current mind/body views that seem to fall into neither the reductive nor the emergent category. Though appearances may mislead us in some cases, as we will see below in section V, there are current options that are generally regarded as in neither family. Most prominently there is the sort of nonreductive physicalism discussed above which is probably the closest to the mainstream view within the philosophic community (Davidson, 1970; Putnam, 1972; 1978; Fodor, 1974; Boyd, 1980; Searle, 1992; Van Gulick, 1992). As noted, it aims to combine some form of ontological physicalism, typically framed in terms of realization, with a rejection of most forms of representational reduction (ONT-Reduction & Not REP-Reduction). To its supporters, this offers the best of both worlds, while its critics (Kim, 1989; Chalmers, 1996) see it as cheating or attempting (unsuccessfully) to pass as a member of the physicalist club without paying one's mandatory dues. Relevant to our current concerns, some of its critics claim it cannot help sliding into an emergent dualism of a particular problematic sort that entails radical causal emergence (Kim, 1992). Were that so, it would indeed be hard-pressed to defend its alleged physicalist status, but it is far from clear that nonreductive physicalism is forced into any such dire commitments.

Current forms of nonreductive physicalism trace back to the mid 1970s and the work of the philosophers Donald Davidson (1970; 1974), Jerry Fodor (1974), Hilary Putnam (1972) and Richard Boyd (1980) among others, which supplanted prior logical empiricist views about the unity of science and the expectation of tight inter-theoretical links and definitions among our various ways of describing and explaining the many levels and aspects of reality. Fodor in particular was concerned to validate what he called the 'autonomy of the special sciences'. Although everything in the world might be ultimately physical, Fodor argued that the nonphysical sciences — be they biology, economics or psychology — provided us with means of describing, explaining, predicting and manipulating the world that were unavailable using the resources of the physical sciences. The old unity of science view (Oppenheim and Putnam, 1958), which held that all true theories must ultimately be translatable into the language of physics, was rejected as a form of conceptual or representational imperialism. Putnam (1978) drew heavily on the pragmatic aspects of representation and explanation and on a practical view of theories as cognitive tools. Whether an all-knowing God might or might not be able to use purely physical descriptions to comprehend all the complexities of the world was of little relevance to our theoretical options as the contextually situated and cognitively limited agents that we are. For all practical purposes, the explanations and understanding available to us through the use of the special sciences are simply not accessible through the use of physical theory alone (Garfinkel, 1981; Van Gulick, 1992). And when one is engaged in the real practice of science, what matters is what is possible in practice by us rather than what an omniscient deity might be capable of doing.

The view obviously has its attraction and fits well with the pluralistic *Zeitgeist* of our time. The challenge has been to show that one can 'eat one's pluralist cake' while still remaining robustly physicalist at the ontological level. Being a nonreductive physicalist myself, I believe the two aspects of the view can be consistently combined, and I have argued for such a view elsewhere (Van Gulick, 1992). However, critics from both left (dualists such as Chalmers, 1996; Hasker,

1999) and right (reductive physicalists such as Kim, 1989) obviously disagree, and debate continues.

Among other recent mind/body options, at least four need to be mentioned: fundamental property dualism, pan (proto-) psychism, dual-aspect monism and what we may call the 'multi-revolutions view'. In the Kuhnian context of 'extraordinary science' these latter views move farther from the normal paradigm and more radically loosen the rules of what might count as a resolution of the mind/body anomaly. Although I list them as four separate categories, the lines between them blur somewhat and just where a specific theory falls is not always clear. With that caveat, let me say just a bit about each.

Fundamental Dualism. All forms of dualism recognize two distinct and separate ontological domains (the mental and the physical). Though contemporary dualism more often appeals to a duality of properties, historically most dualists were committed to a duality of substances and the latter view still has present supporters.

We have already discussed emergent property dualists. Though they treat mental properties as distinct from physical ones, they do not regard mental properties as fundamental. On the emergent property dualist view, mental properties are something over and above their physical bases, but they are not fundamental in so far as they owe their existence to their emergence from their nonmental physical bases. They are *more than* their bases, but they are nonetheless in some way *dependent upon* them. By contrast, fundamental property dualism gives basic mental properties the same bedrock foundational status as our fundamental physical forces (Chalmers, 1996). Just as the nuclear force and the electromagnetic force are equally primitive features of physical reality according to the so-called standard theory, so too are basic mental and physical properties according to the fundamental property dualist. Thus, any lawlike links between them would be rock-bottom as well and not open to explanation in terms of more basic underlying regularities. If so, we would expect mind/body explanation to terminate at a core of nomically primitive psycho–physical laws, which we would just have to accept as the way things work in our world: 'No more questions please!'

It is certainly a consistent option within the space of possible views, and not open to *a priori* refutation. Yet it seems explanatorily unsatisfying and empirically at odds with our evidence from other cases about how properties at different levels of our world relate (McGinn, 1991). If chemical properties, biological properties, and economic properties are not fundamental, does that not give us at least some good reason for believing that mental properties are probably not either? The property dualist nonetheless believes that there are good reasons for regarding mental properties as a special case (Chalmers, 1996) and debate continues.

Fundamental substance dualists regard minds as independently existing substances or things, distinct from any physical systems with which they may interact. Classical Cartesian dualism treated minds and bodies as distinct substances with their own respective essences. Both popular (i.e., everyday) dualism and traditional religious (Christian) dualism regard souls or minds as distinct non-physical substances. In recent years the philosophers John Foster and Richard Swinburne have both argued for neo-Cartesian forms of dualism. Their arguments proceed from diverse premises. Foster (1991) relies essentially upon the supposed inadequacy of all forms of reductionism to establish the nonphysicality of mental properties. He goes on to argue that nothing other than a wholly nonphysical subject could be the bearer of such irreducible intrinsic mental properties. In effect, he argues from property dualism to substance dualism. Swinburne (1986) appeals to the necessity of a soul to provide an adequate basis for personal identity through time and change, especially for what he regards as the coherent possibility of the persistence and continuity of the self through the loss of one's body and the conceivability of disembodied existence.

All forms of fundamental dualism — whether property or substance — must confront the causal interaction problem that has bedevilled dualism at least since the days of Descartes' critics in the seventeenth century. We seem to have abundant observational evidence of two-way causal interactions between minds and bodies. But if mental properties are fundamentally distinct from the physical, then they could not affect the course of physical events without violating the causal closure of the physical (Kim, 1990; 1999). Thus, the fundamental property dualist seems forced to chose between two unsatisfactory alternatives: giving up the causal closure of the physical or regarding mental properties as epiphenomenal (at least qua the physical domain) despite the enormous body of evidence that appears to show otherwise. Physicalists understandably find neither option attractive. However, many dualists are willing to deny the causal closure of the physical. However much the view may conflict with our mainstream views about the causal closure of the physical world, future evidence could empirically support some form of fundamental dualism, and so it stays on the table as a possibility, whether likely or not.

Pan (Proto-) Psychism. If it is impossible, as some claim, to build a mind or consciousness out of purely nonmental parts, then perhaps we must view the parts as themselves in some way having a mental or at least proto-mental aspect (Nagel, 1979; Chalmers, 1996). That is just what pan (proto-) psychics claim; everything real has a mental (or proto-mental) aspect down to the smallest units of physical reality, though it is difficult to understand in what sense an atom or electron could have features that were even in the most remote sense mental. Arguments for the view are typically indirect and rely on the assumption that conscious minds cannot be constructed out of totally non-mental parts. However, panpsychism, proto or otherwise, is generally very short on details about how mentality might be universally present through out physical reality. The idea of proto-psychic features also confronts a dilemma: the more we view them as *like* familiar mental properties, the more implausible it is to suppose that they could be universally present in simple physical components — how could a molecule be aware, or how could an atom experience red or feel pain? But, conversely, the more we view them as *unlike* familiar mental features, i.e., the more we emphasize their 'proto-ness', the more difficult it is to see how they might give rise to consciousness. The basic explanatory gap just reoccurs at the boundary between the proto-psychic and the conscious. The viability of the view depends upon giving some plausible positive account of how simple physical components could have mental aspects of a sort apt for producing consciousness, and as yet no such account has been given.

Dual-Aspect Monism. Closely related to the pan-psychism but worth distinguishing, is dual-aspect theory or so-called neutral monism, which asserts that reality is ultimately constituted by a single realm of things and properties (hence the monism) which are neither mental nor physical. This ultimate ('ur') reality manifests itself to us in both physical and mental ways, but is itself more basic than either. Such views might be found historically in Spinoza or Leibniz's theory of monads. Early in the twentieth century, it was advanced on by Bertrand Russell (1921) and various logical empiricists (Schlick, Ayer), and the view still has supporters today (see Strawson, 1994, though he explicitly disavows the 'dual aspect' label in favour of 'agnostic materialism').

Multi-Revolutions View. One of the more extreme responses to the supposed psycho–physical anomaly has been to call for more than one revolution. Proponents of this view, including the physicist Roger Penrose (1989; 1994) and the philosophers Michael Lockwood (1989) and Colin McGinn (1995), have argued that the persistently mysterious nature of the psycho–physical gap gives good reason to believe that we need new ways of conceptualizing and understanding both the nature of the mental and the nature of the physical. Our inability to solve the puzzle of their link results, they say, from the inadequacy of both sides of the equation. McGinn, for example, claims that explaining the link would require both a better understanding of what he calls the hidden nature of consciousness (1991) and a radically different conception of physical space (1995). It is because he doubts that we humans are capable of forming the requisite novel concepts that he takes such a pessimistic view of our human prospects of resolving the anomaly. Lockwood (1989) finds the concept of matter itself deeply problematic and argues for what might be regarded as a dual-aspect view in which matter and mind are more closely integrated at the fundamental level. Penrose finds existing attempts to explain consciousness in terms of physical or algorithmic processes doomed to failure for reasons concerned with the mathematical limits of formal systems; he is equally dissatisfied with the present attempts to integrate our physical theories of the very small and the very large at the interface of quantum mechanics and general relativity. He optimistically hopes for a joint revolution that would address and resolve both puzzles. Some critics regard the search for a 'two-for-one' solution as little more than wishful thinking, but it is one of the features of extraordinary science that people begin to entertain more and more radical alternatives, and sometimes those extreme directions provide the solution, though not often.

A variant of the two-revolutions view argues that the second revolution is needed not in physics but merely in what they regard as the outdated conception of physics held by most philosophers. For example, Michael Silberstein (1999) argues that physics, and especially quantum mechanics, has already rejected the sort of atomistic and localist view assumed by most philosophic parties to the mind/body debate. Both physicalists and their dualist opponents still typically think of physical explanations in a localist way that requires that all system-level properties be explained as consequences of the properties of the system's parts and mode of combination. Silberstein argues to the contrary that in quantum mechanics, as demonstrated by cases such as those involving entangled particles, systems are regarded as having properties over and above those determined by their parts. One might be a physicalist while rejecting standard atomism. If physics has, in fact, already accepted such a view, which might be regarded as a form of radical metaphysical emergence, then the constraints that limit what can count as a physically acceptable solution to the mind/body problem may indeed be very different from what most philosophers typically assume.

Having completed the survey promised in my title, we can pull all the mind/body options we have discussed into a single diagram which hopefully can serve as a useful pocket map for those travelling through the philosophic terrain. Figures 3 and 7 can be thought of as insets that show more local details in the two ten-member families listed first.

CURRENT OPTIONS ON THE MIND/BODY PROBLEM

- **Reduction**: 10 versions (see *Figure 3*)

- **Emergence**: 10 versions (see *Figure 7*)

- **Other Options**:

    Mainstream: Nonreductive physicalism

    More radical: Fundamental Dualism
    - Property
    - Substance

    Pan (Proto-) Psychism

    Dual Aspect Monism

    Multi-Revolutions View

Figure 8

V: Selected Conflicts, Agreements and Other Relations

Having distinguished more than two dozen different options, we can in this final section consider only a few of their possible interrelations (there being 26 x 25 possible two-way cross pairings alone, leaving aside combinations involving more than two views). Since my intent has been to act more as a tour guide rather than an advocate, I will offer just five general observations aimed at further clarifying the lie of the philosophic landscape. It may contain some paths or links between positions other than those that are normally assumed, and being aware of them should help in navigating the terrain.

With that intent, I offer the following five thoughts on the map and the space of views it aims to depict.

1. *Pay attention to the 'key' — the need for clarity and the avoidance of conflation.* Labels and terms are used with such a diversity of meanings in the mind/body literature that it is absolutely essential that one be clear about what meaning is intended on a given use. Because central terms, such as 'reduction' and 'emergence' have no standard interpretation, one should always make one's own use clear; provide a 'key' for reading the verbal map of your view. And take care in reading others to hear their words as they intended them; otherwise it's all too easy to mislocate them and for the discussion to get lost in a fog of misunderstanding. The general need for clarity is platitudinously obvious, but it's so important in this context that it merits restating nonetheless.

2. *Respect the subjective/objective division.* Perhaps nowhere is the need for clarity greater or the threat of conflation more likely than when one is dealing with subjective and objective versions of some notion. As we saw above, the subjective/objective distinction runs through the space of options dividing both the reduction and emergence families into distinct and separate sections, just as a mountain range might separate a chain of plains or valleys through which it runs.

The reduction region divides into objective relations of ontological reduction such as identity, composition, or realization, and subjective regions of representationally reductive relations such as derivability, conceptual necessitation, or pragmatic equivalence. The parallel division within the emergence region is between objective notions of metaphysical emergence such as modest or radical kind emergence, and subjective notions of epistemic emergence concerning the limits on our cognitive abilities to explain or understand the features of wholes in terms of those of their parts. (N.B. Using the labels 'objective' and 'subjective' here is meant only to draw a distinction between relations among things in the world [objective] and relations among our ways of thinking about or representing things in the world [subjective]. Although those same words are used in many ways in the literature, other meanings should not be read into their use here. In particular, there is no intent to use 'subjective' to imply restriction to a first-person experiential point of view, as when Thomas Nagel classifies facts about experience as subjective because he regards them as fully understandable only from the perspective of those able to have similar experiences themselves. See Nagel, 1974; 1986; Van Gulick, 1985; Lycan, 1990.)

The most common and controversial moves with respect to this division concern attempts to reach ontological conclusions from subjective premises. Facts about our human incapacity to reductively explain how consciousness might be realized by underlying physical processes cannot by themselves justify us in concluding that consciousness is not a physically realized process. The explanatory gaps may reflect subjective limits on our (current) human conceptual or imaginative capacities rather than any objective divisions in the world. Conceivability arguments for dualism — whether offered by Descartes (1642), Saul Kripke (1972) or more recently by David Chalmers (1996) — are regarded by physicalists as tripping on this mistake. The apparent possibility to conceive of worlds that contain molecule-for-molecule physical duplicates of humans lacking any conscious mental life does not entitle the dualist to conclude that conscious properties are not physical in any ontologically robust sense (e.g., identity or realization). According to the physicalist critics, making such an inference would require us to pass invalidly across the subjective/objective divide, moving from facts about the limits of our concepts to an objective claim about the distinctness and independence of the real-world features to which we refer by use of those concepts. The dualist making such an inference would need to show that the concepts he employed on both the mental and physical sides of his thought experiment were adequate to support such a metaphysical conclusion (Van Gulick, 1999); physicalists doubt the dualist can discharge that burden. In a Shakespearean parody, 'the fault may lie not in the world but in our concepts of it'.

Thus, we should not infer that mental properties are ontologically nonphysical just because we cannot representationally reduce our mental concepts or theories to physical ones. Nor should we conclude that mental properties or powers are metaphysically emergent just because they are subjectively emergent relative to our abilities to explain, predict or understand using resources of physical theory. Additional argument is needed to justify the move from subjective premises to objective conclusion. The dualist champions of conceivability arguments believe additions can be made that validate the move (Chalmers, 1996), but physicalist critics argue to the contrary (Yablo, 1999; Van Gulick, 1999).

3. *Reduction and emergence can overlap (not necessarily disjoint).* Although the notions of reduction and emergence are often paired as polar opposites, there are, in fact, many consistent combinations of views from the two respective families. The slogans associated with the two make them seem mutually exclusive. How could Xs 'just be Ys' or 'merely special sorts of Ys' but also be 'something other than Ys'? Or how could Xs be 'something over and above Ys' but also 'nothing more than Ys'? The contradictions seem immediate, and indeed they are if one assumes a consistent reading for both conjuncts. But as we saw above there are many versions of reduction and perhaps equally as many versions of emergence; we distinguished at least ten of each.

Though some versions of reduction are strictly inconsistent with some versions of emergence, other cross combinations involve no necessary conflict. One could not consistently combine an identity version of ontological reduction with a metaphysical notion of radically kind emergent properties or powers. But one might without contradiction accept both a cognitive/explanatory emergence view of mental properties as well as a realization or composition version of ontological reduction.

4. *Nonreductive physicalism lies largely within the intersection of the dual families of emergence and reduction, rather than in a third and separate region.* Nonreductive physicalism is typically regarded as an option wholly distinct from

either reductive physicalism or emergence. However, once one recognizes that reduction and emergence are not mutually exclusive in all their versions, one can see that nonreductive physicalism actually occupies a region within the intersection of the reduction and emergence families. Those two families are to some extent complementaries or duals of each other, especially *vis-à-vis* the subjective/objective division. Thus, nonreductive physicalism on one hand combines a denial of (subjective) representational reduction with an acceptance of some robust form of (objective) ontological reduction such as physical realization. On the other hand, it pairs the denial of at least the strongest forms of ontological emergence such as radical kind emergence with an acceptance of epistemic emergence in either or both of its forms. Put symbolically:

Nonreductive Physicalism => (Ont Reduction & Not REP-Reduction).

Nonreductive Physicalism => (Epistemic Emergence & Not Radical Metaphysical Emergence).

Despite the 'non' in the name of their position, nonreductive physicalists do accept some forms of objective (i.e., ontological) reduction, while rejecting most forms of subjective (i.e., representational) reduction. In the dual domain of emergence, they take the complementary position. They accept various forms of subjective (i.e., epistemic) emergence but reject the radical versions of objective (i.e., metaphysical) emergence as shown in figure 9.

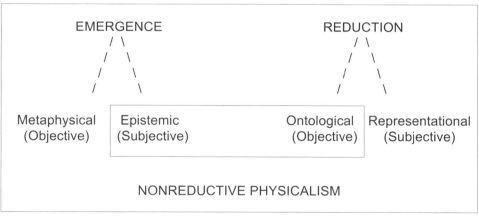

Figure 9

Thus, contrary to common belief, our map of logical space should not locate nonreductive physicalism in a region disjoint and separate from those occupied by the emergence and reduction families, but rather in a special sub-region of their intersection.

5. *What looks like a gap (an anomaly or a crisis) depends on your viewpoint and location.* As noted in the Kuhnian introduction to this paper, much of the pressure for extraordinary philosophical theorizing about consciousness and the psycho–physical link arises from the sense of a persistent anomaly that resists satisfactory resolution by more mainstream physicalist approaches.

There are, no doubt, other factors at work as well. Our current scientific knowledge about the psycho–physical link is admittedly primitive, and there are at present no plausible detailed empirical models of how consciousness might arise from a wholly physical substrate. Research has not advanced far enough to generate a consensus paradigm around which a body of 'normal scientific' practice might coalesce. Extraordinary theorizing is the rule at such early stages, as multiple investigators strike off in differing directions in search of some means to gain significant initial progress on the problem. Such an unsettled scientific state might by itself elicit a good deal of extraordinary philosophic speculation, even if the philosophic community remained largely committed to some form of mainstream physicalism as its own normal practice paradigm. The impulse to engage in extraordinary theorizing is likely contagious and easily spread from one domain to another.

Moreover, extraordinary theorizing need not have any specific trigger at all. Paul Feyerabend, the iconoclastic philosopher of science and champion of revolutionary science, argued that extraordinary science is not and should not be restricted to periods of crisis and anomaly (Feyerabend, 1975). We need not wait until prevailing research programmes 'break' or fail to make progress before exploring radically alternative approaches to the field. Feyerabend had no sympathy with an 'If it's not broke, don't try to fix it' approach to science. He believed that some level of extraordinary or revolutionary science should be going on at all times. Perhaps a good alternative slogan for the Feyerabendian view might be 'Even if it's not broke, you might find something better', or 'If you don't start looking till it is broke, you're not likely to have anything to put in its place when it breaks'. Thus, there are at least two possible sources for extraordinary philosophic theorizing about the psycho–physical link other than the supposed perception of a recalcitrant anomaly. Nonetheless, extraordinary activity is far more likely to occur in contexts of anomaly and perceived crisis, and such a sense does seem to exert a lot of pressure on the current state of philosophic play on the mind/body issue.

But is the perception correct or well founded? Is mainstream physicalism in fact in a crisis caused by its persistent inability to resolve a problem or puzzle that falls clearly within its domain? The answer is not as obvious as it might at first seem. Surely there are puzzles it has not solved and questions to which it has given no satisfying answers, but not every failure to solve a problem counts as an anomaly. As Kuhn (1962) made clear from the start, what counts as a problem or its solution is determined on the whole internally by the research community itself through its paradigms of normal practice. Unless a puzzle presents a problem of a sort that the community by its own internal standards counts as one it ought to be able to solve, its failure to do so need not generate any anomaly or sense of crisis. And even if a puzzle meets the standards to count as a valid problem, it need not be open to a quick or easy solution. If so, a string of negative attempts need not indicate anything more than the difficulty of the problem and the need for continuing effort from within the paradigm.

Physicalism's supporters and its critics will obviously differ in their assessments of its current status. Its critics will argue that it is no closer to explaining how consciousness might be a physical process than it was three hundred and fifty years ago when Descartes and Leibniz expressed their early scepticism. Nor is it just the explanatory details that we lack. As Thomas Nagel pointed out more then twenty-five years ago in his famous 'What is it like to be a bat?' article (1974), we have a model of how to begin to bring the two sides of the psycho–physical divide together; physicalists still find themselves staring at an explanatory blank wall. Solutions to hard problems rarely come quickly, but if three centuries of failure do not suffice to generate a crisis, what more is needed? Physicalists in reply might compare their inability at the end of the twentieth century to explain the physical nature of consciousness with that of their predecessors at the start of the century to do the same for life. Vitalism was still a serious scientific position one hundred years ago, and a sense of mystery and bafflement still attended attempts to explain life, growth and reproduction as wholly physical. It was only at mid-century that the puzzle was solved when the modern biochemical revolution provided us with more adequate concepts of both the biological processes and their physical substrates. Only when we had a better understanding of both sides of the equation could we see how they fit together. Optimistic physicalists hope that the twenty-first century will do the same for the psycho–physical link. Predictions about what sorts of explanations the future will provide are typically based as much on hope and prior outlook as on evidence, and at present there is no way to decide between optimistic physicalists or their pessimistic critics.

However, it is worth noting that on the Kuhnian model, crises get resolved in many ways, and not all resolutions involve a solution to the problem. They end in at least four common ways:

(1) Through a revolution which solves or rejects the resistant problem.

(2) Through an internal solution of the problem.

(3) Through postponement; putting the problem off for later solution when the field is more advanced.

(4) Through an internal dissolution/rejection of the problem.

How then should we think about the alleged failure to solve the mind/body problem, especially the supposed failure of physicalists to provide any adequate solution to the consciousness/brain version of the problem? There are no doubt major questions as yet unanswered about that relation. At an abstract level, physicalists may claim that consciousness is realized by a wholly physical substrate, but they cannot at this point offer an intuitively satisfying story about how such a realization account might work. Constructing such a model may be more a scientific job than a philosophical one, but in the absence of any such concrete model, abstract physicalist claims about realization ring a little hollow. Perhaps solutions can be found more-or-less within the normal physicalist paradigm, either in the near term or later at a more advanced stage of theorizing, but perhaps not. The solution may instead require a revolution in our ways of thinking about mind, matter, or both. We cannot at this point predict with any great confidence whether the apparent crisis will end in a type 1, 2, or 3 way.

However, there are some explanatory demands that physicalists, at least those accepting some form of mainstream nonreductive physicalism, can justifiably reject. In particular, they should resist the claim that solving the mind/body problem requires deriving our mentalistic theories from our physicalist ones, as reductive logical empiricists had argued. Nor should they agree that doing so must provide us with reductive explanations that allow to see on purely *a priori* conceptual grounds how the physical facts necessitate their conscious consequents, as the dualist proponents of conceivability arguments assume.

Given a pragmatic and contextual account of representation and explanation of the sort that nonreductive physicalists typically hold for quite independent and general reasons, there are good reasons to reject the logical empiricist and apriorist views of what an adequate solution requires. Thus, failures to solve the mind/body problem relative to those representationally reductive criteria need not embarrass the nonreductive physicalist nor count as any sort of anomaly from her perspective. Alleged crises based on our inabilities to provide such reductive accounts might end through dissolution of type 4 rather than a solution; the physicalist may reject the demand for any such account as illegitimate.

Just like other normal practice problem solvers, physicalists get a say in what counts as a problem and a solution within their domain, especially when those views are not *ad hoc* or merely self-protective but are instead based on independently motivated claims about the nature of explanation. Of course, having an internal say in what counts as a solution does not in itself guarantee that outsiders will or should share your evaluation, any more than most of us would be swayed by an astrologer's self-proclaimed success in solving astrological problems in astrologically valid ways. Mainstream physicalists have a responsibility to support and defend their criteria for solving the mind/body problem, but they have plenty of resources for doing so, and their critics have an equal obligation to defend their more reductive standards.

Thus, an accurate map of field should reflect the fact that which problems count as solved, unsolved or illegitimate depends upon one's own location within the space of possible positions. What links one thinks can or cannot be seen will turn in part upon the lie the land appears to have from where one stands. Nor need there be an outside point of view that shows with neutral truth what links there are; there may be only inside answers to the question.

References

Beckermann, A., Flohr, H. and Kim, J. (ed. 1992), *Emergence or Reduction? Essays on the Prospects for Nonreductive Physicalism* (Berlin: DeGruyter).

Block, N. and Stalnaker, R. (1999), 'Conceptual analysis, dualism, and the explanatory gap', *Philosophical Review*, **108**, pp.1–46.

Boyd, R. (1980), 'Materialism with reductionism: what physicalism does not entail', in *Readings in Philosophy of Psychology Volume 1*, ed. N. Block (Cambridge, MA: Harvard University Press).

Chalmers, D. (1996), *The Conscious Mind* (Oxford: Oxford University Press).

Churchland, P.M. (1981), 'Eliminative materialism and the propositional attitudes', *Journal of Philosophy*, **78**, pp. 67–90.

Churchland, P.M. (1985), 'Reduction, qualia, and the direct introspection of brain states, *Journal of Philosophy*, **82**, pp. 8–28.

Davidson, D. (1970), ' Mental events', in *Experience and Theory*, ed. L. Foster and J. Swanson (Amherst: University of Massachusetts Press). Reprinted in Davidson (1980).
Davidson, D. (1974), 'Psychology as philosophy' in *Philosophy of Psychology*, ed. S.C. Brown (New York: MacMilllan). Reprinted in Davidson (1980).
Davidson D. (1980), *Essays on Actions and Events* (Oxford: Oxford University Press).
Dennett, D. (1988), 'Quining qualia', in Marcel and Bisiach (1988).
Descartes, R. (1642), *Meditations on First Philosophy* (Paris)
Feigl, H. (1958), ' The mental and the physical', in *Minnesota Studies in the Philosophy of Science Volume II*, ed. H. Feigl, M. Scriven and G. Maxwell (Minneapolis: University of Minnesota Press).
Feyerabend, P. (1975), *Against Method* (London: New Left Books).
Fodor, J. (1974), 'Special sciences, or the disunity of science as a working hypothesis', *Synthese* , **28**, pp. 77–115. Reprinted in Fodor (1981).
Fodor, J. (1981), *Representations* (Cambridge, MA: MIT Press).
Foster, J. (1991), *The Immaterial Self* (London: Routledge).
Frege, G. (1892), 'Über Sinn und Bedeutung', *Zeitschrift für Philosophie und Philosophische Kritik*, **100**, pp. 25–50. Translated by Max Black (1952) as 'On sense and reference', in *Translations from the Philosophical Writings of Gottlob Frege*, ed. P. Geach and M. Black (Oxford: Oxford University Press).
Garfinkel, A. (1981), *Forms of Explanation* (New Haven, CT: Yale University Press).
Hasker, W. (1999), *The Emergent Self* (Ithaca: Cornell University Press).
Hill, C. and McLaughlin, B. (1999), 'There is less in reality than dreamt of in Chalmers' philosophy', *Philosophy and Phenomneological Research*, **59**, pp. 445–54
Jackson, F. (1982), 'Epiphenomenal qualia', *Philosophical Quarterly*, **32**, pp.127–36.
Jackson, F. (1986), 'What Mary didn't know', *Journal of Philosophy*, **32**, pp. 291–5.
Jackson, F. (1998), 'Postscript on qualia,' in *Mind, Method and Conditionals,* ed. F. Jackson (London: Routledge).
Kemeny, J. and Oppenheim, P. (1956), 'On reduction', *Philosophical Studies*, 7, pp. 6–17.
Kim, J. (1982), 'Psychophysical supervenience', *Philosophical Studies*, **41**, pp. 51–70.
Kim, J. (1989), 'The myth of nonreductive physicalism', *Proceedings and Addresses of the American Philosophical Association*, **63**, pp. 31–47. Reprinted in Kim (1993a).
Kim, J. (1990), 'Explanatory exclusion, and the problem of mental causation', in I*nformation, Semantics and Epistemology*, ed. E. Villanueva (Oxford: Basil Blackwell).
Kim, J. (1992), '"Downward causation" and emergence', in *Emergence or Reduction? Essays on the Prospects for Nonreductive Physicalism*, ed. A. Beckermann, H. Flohr and J. Kim (Berlin: DeGruyter).
Kim, J. (1993a), *Supervenience and Mind* (Cambridge: Cambridge University Press).
Kim, J. (1993b), 'Multiple realization and the metaphysics of reduction', in *Supervenience and Mind*, ed. J. Kim (Cambridge: Cambridge University Press).
Kim, J. (1999), *Mind in a Physical World* (Cambridge, MA: MIT Press).
Kripke, S. (1972), *Naming and Necessity* (Cambridge, MA: Harvard University Press).
Kuhn, T. (1962), *The Structure of Scientific Revolutions* (Chicago: University of Chicago Press).
Levin, J. (1986), 'Could love be like a heat wave? Physicalism and the subjective character of experience', *Philosophical Studies*, **49**, pp. 245–61.
Levine, J. (1983), 'Materialism and qualia: the explanatory gap', *Pacific Philosophical Quarterly*, **64**, pp. 354–61.
Levine, J. (1993), 'On leaving out what it's like', in *Consciousness*, ed. M. Davies and G. Humphreys (Oxford: Blackwell).
Lewis, D (1982), 'Postscript to mad pain and Martian pain', in *Philosophical Papers Volume I*, ed. D. Lewis (Oxford: Oxford University Press).
Loar, B. (1990), 'Phenomenal states', in *Philosophical Perspectives Volume 4: Action Theory and the Philosophy of Mind*, ed. J. Tomberlin (Atascadero, CA: Ridgeview Publishing).
Locke, J. (1688), *An Essay on Human Understanding* (London).
Lockwood, M. (1989), *Mind, Brain and Quantum* (Oxford: Blackwell).
Lycan, W. (1990), 'What is the "subjectivity" of the mental?', in *Philosophical Perspectives Volume 4: Action Theory and the Philosophy of Mind*, ed. J. Tomberlin (Atascadero, CA: Ridgeview Publishing).
Marcel, A. and Bisiach E. (1988), *Consciousness in Contemporary Science* (Oxford: Clarendon Press).
McGinn, Colin (1989), 'Can we solve the mind–body problem?', *Mind* , **98**, pp. 349–66.
McGinn, Colin (1991), *The Problem of Consciousness* (Oxford: Blackwell).
McGinn, Colin (1995), 'Consciousness and space', in *Conscious Experience*, ed. T. Metzinger (Thorverton: Imprint Academic).
Moore, G.E. (1902), *Principia Ethica* (Cambridge: Cambridge University Press).
Nagel, E. (1961), *The Structure of Science* (New York: Harcourt, Brace and World).
Nagel, T. (1974), 'What is it like to be a bat?', *Philosophical Review,* Reprinted in *Mortal Questions*, ed. T. Nagel (Cambridge: Cambridge University Press).
Nagel, T. (1979), 'Panpsychism', in *Mortal Questions,* ed. T. Nagel (Cambridge: Cambridge University Press).
Nagel, T. (1986), *The View from Nowhere* (Oxford: Oxford University Press).

Oppenheim, P. and Putnam, H. (1958), 'Unity of science as a working hypothesis', in *Minnesota Studies in the Philosophy of Science Volume II*, ed. H. Feigl, M. Scriven and G. Maxwell (Minneapolis: University of Minnesota Press).
Penrose, R. (1989), *The Emperor's New Mind* (Oxford: Oxford University Press).
Penrose, R. (1994), *Shadows of the Mind* (Oxford: Oxford University Press).
Place, U.T. (1956), 'Is consciousness a brain process?', *British Journal of Psychology*, **47**, pp. 44–50.
Putnam, H. (1972), 'Philosophy and our mental life', in *Mind, Language and Reality, Philosophical Papers Volume 2*, ed. H. Putnam (London: Cambridge University Press).
Putnam, H. (1978), *Meaning and the Moral Sciences* (London: Routledge and Kegan Paul).
Rorty, R. (1970), 'In defense of eliminative materialism', *The Review of Metaphysics*, **24**, pp.112–21.
Russell, B. (1921), *The Analysis of Mind* (London: Macmillan).
Searle, John (1992), *The Rediscovery of the Mind* (Cambridge, MA: MIT Press).
Silberstein, M. (1999), 'Emergence and the mind–body problem', *Journal of Consciousness Studies*, **5** (4), pp. 464–82.
Silberstein, M. and McGeever, J. (1999), The search for ontological emergence', *The Philosophical Quarterly*, **49**, pp. 182–200.
Smart, J.J.C. (1959), 'Sensations and brain processes', *Philosophical Review*, **82**, pp. 141–56.
Sperry, R. (1983), *Science and Moral Priority: Merging Mind, Brain and Human Values* (New York: Columbia University Press).
Sperry, R. (1991), 'In defense of materialism and emergent interaction', *Journal of Mind and Behavior*, **12**, pp. 221–45.
Strawson, G. (1994), *Mental Reality* (Cambridge, MA: MIT Press).
Swinburne, R. (1986), *The Evolution of the Soul* (Oxford: Oxford University Press).
Van Gulick, R. (1985), 'Physicalism and the subjectivity of the mental', *Philosophical Topics*, **12**, pp. 51–70.
Van Gulick, R. (1992), 'Nonreductive materialism and intertheoretical constraint', in Beckermann *et al.* (1992).
Van Gulick, R. (1993a), 'Understanding the phenomenal mind. Are we all just armadillos?', in *Consciousness*, ed. M. Davies and G. Humphreys (Oxford: Blackwell).
Van Gulick, R. (1993b), 'Who's in charge here and who's doing all the work?', in *Mental Causation*, ed. J. Heil and A. Mele (Oxford: Clarendon Press).
Van Gulick, R. (1999), 'Conceiving beyond our means: the limits of thought experiments', in *Toward a Science of Consciousness III*, ed. S. Hameroff, A. Kazniak and D. Chalmers (Cambridge MA: MIT Press).
Wilkes, K. (1988), '—, yishi, duh, um and consciousness', in Marcel and Bisiach (1988).
Wilkes, K. (1995), 'Losing consciousness', in *Conscious Experience*, ed. T. Metzinger (Thorverton: Imprint Academic).
Yablo, S. (1999), 'Concepts and consciousness', *Philosophy and Phenomenological Research*, **59**, pp. 455–63.

Harry T. Hunt

Some Perils of Quantum Consciousness

Epistemological Pan-experientialism and
the Emergence–Submergence of Consciousness[1]

If consciousness emerges into ontological reality at some point in nature, as system complexity increases, then it also 'submerges' at some adjoining point, as structures simplify. This has led some to posit a 'latent-consciousness' in what Bohr saw as the consciousness-like spontaneity of quantum phenomena. Yet to move on this basis to Whitehead's ontological pan-experientialism or to direct quantum explanations of consciousness (Hameroff and Penrose) faces serious epistemological limitations — perhaps being more unwittingly projective than genuinely explanatory. More reasonable would be an epistemological pan-experientialism in the sense of the later James. Consciousness, as the ultimate lens and medium of all knowledge, is inseparable from the physical reality it would know, especially at the very limits of empirical observation in microphysics. 'Submerged' consciousness is better understood in Jamesian pragmatic terms than via assumed but unprovable ontologies.

Approaches to consciousness in terms of quantum microphysics have often linked quantum field effects and states of consciousness in terms of their ostensibly shared features of holism, spontaneity, complementarity, and observational indeterminism (King, 1997; Rosenblum and Kuttner, 1999). However, these links can be understood either in terms of an ontology of what is real, or, more conservatively, in terms of an epistemology of what is knowable. If we take them as real, then a kind of proto-consciousness is being posited as part of quantum effects. This can be understood either as a reductionist explanation of consciousness itself, as in the original Hameroff-Penrose microtubule hypothesis, wherein consciousness is produced by quantum superimpositions in neuronal microtubules (Hameroff, 1994; Penrose, 1994), or in terms of a literal philosophical pan-

[1] A preliminary version of this paper was presented at the conference 'Tucson 2000: Toward a Science of Consciousness', April 14, 2000.

experientialism as posited by Whitehead (1929), and modified by Chalmers (1996) and Hameroff. These alternatives are surprisingly difficult to tell apart, as we shall see below, which is also consistent with the more purely epistemological solution proposed herein: namely, that locating consciousness features at the quantum level is based on confusing the limitations of observation, common to both microphysics and consciousness, with an ontology of their relation at precisely the point where these become indistinguishable.

I will argue that the most parsimonious way forward here is to posit an emergentist solution, in which consciousness appears in the universe as the most hierarchically complex system we know, or possibly could know, and which re-creates, on its new emergent level, principles first manifested on quantum and nonlinear systems levels. It would be the parsimony of nature to re-use the same basic principles on very different levels of complexity, as in the theoretical complementarity of a simultaneous flow and particulate structure for both light and consciousness, and in the methodological indeterminism by which observation changes what is being observed — again, in both microphysics and introspected consciousness. That the same organizing principles would appear on levels of reality so widely separated by other levels of analysis not requiring these principles (biological, molecular and atomic) makes it unlikely, on this view, that one could 'explain' the other (Scott, 1995) — either as microphysics explaining consciousness or a pan-experiential proto-consciousness somehow explaining quantum phenomena.

Sperry (1987), developing the perspective of a systems analysis, suggested that consciousness needed to be considered as a 'greater than the sum of its parts' emergence or gestalt. Once emerged, consciousness would constitute its own level of analysis with its own requirements, in much the same way that chemistry, while resting on physics, has its own methodology and principles that are pursued largely independent, with important selective exceptions, of the principles of microphysics. However, if we are to speak, with Sperry, of consciousness having a point of 'emergence', then logically we should also be able to ask about its 'submergence' or latency. Indeed, we do find separate 'features' of consciousness, such as relative spontaneity, holism, complementarity and indeterminism, multiply distributed through less complex natural systems. But consciousness *features* do not a consciousness make, even if they are in the microtubules (Hameroff) and water dipoles (Jibu and Yasue, 1997) of neurons. Instead, emergentist, pan-experientialist, and quantum accounts of consciousness can all be seen as variations within what we could loosely call a top-down science of synthesis, in that instead of analytically isolating the simple within the complex, these locate the seeds of complexity within the simple.

We will see in more detail below that it seems safer here to follow the approach of the later William James (1912) and not Whitehead himself, and so interpret the presence of consciousness-like features on a quantum level in terms of a pan-experientialism that is epistemological, but not provable as ontology. 'Pure experience' in James' phenomenological sense is the necessary lens and medium for all human knowledge, and so it cannot be fully separated from the physical universe it would know and of which it is also the most hierarchically complex expression. So we will have no choice but to see our own consciousness in certain consciousness-like features of physical reality, while not confusing these features with its causation.[2]

An Ontology of Emergence and Submergence

What we can say, without undue assumptions, is that at *some* point systems meeting the full criteria for consciousness do appear as part of reality and that their presence and principles of organization cannot be *ultimately* foreign to the universe that brings them forth over immense spans of time — as in the so-called weak anthropic principle (Hawking, 1988). But this is a far cry from finding a form of consciousness itself in quantum fields or cell microtubules. Niels Bohr (1934), one of the key formulators of quantum theory, has in common with Whitehead, and now Hameroff and Penrose, that they search for a proto-consciousness in physical nature by going 'inward'. They locate, respectively, an inner spontaneity, micro-duration, or uncollapsed wave-packet as hidden *inside* a physical husk of some kind — ostensibly on the notion that consciousness is a something 'within', in contrast to a third-person behaviour that somehow lacks consciousness as a something 'without' or 'outside' an inner conscious core. But this follows only on increasingly questionable definitions of consciousness that make it 'private' and 'hidden'. I would call this 'Whitehead's fallacy', in that from a more functional evolutionary perspective an inner 'spontaneity' can only be conscious if it has a true surrounding environment it is moving within. Indeed, there is some agreement that simple awareness or sentience appears only when organisms move sufficiently in relation to their surround such that an ongoing, sensitive self-location is needed for survival (Miller, 1981; Hunt, 1995; Sheets-Johnson, 1998). It would be just such movement that would be the primary empirical indicator of such an 'inner' awareness. Without the right criteria for the presence of consciousness, its point of emergence in the natural order will be misplaced or misleading.

Meanwhile, on the human level of consciousness, the 'privacy' model is a special limiting case of a more primary definition of consciousness as social and intersubjective. Specifically, the primary definition of consciousness *both* historically and in the dictionary is as a shared or joint awareness, with *both* first- and third-person criteria or modes of access (Natsoulas, 1983). The privacy definition is more metaphysical, perhaps a cultural artefact of our hyper-emphasis on autonomy and individuality (see Hunt, 1995). Now a key point here is that the third-person perspective is not a purely physical one, like that of physics or physiology, but refers to our access to the experience of other sentient beings on the basis of our direct recognition of their behavioral expressions. The pragmatic-descriptive approach of the philosopher Wittgenstein is of help here:

---

[2] This is not to deny the broader value of Whitehead for a science of synthesis and systems theory, in that he is using a *language* of consciousness to locate these actual nonconscious seeds of a higher emergent complexity.

> My thoughts are not hidden from him but are just open to him in a different way than they are to me (Wittgenstein, 1992, p. 34–5).
>
> Consciousness is as clear in his face and behavior as in myself (Wittgenstein, 1980, p. 164).

As Wittgenstein says, an *inner* state always stands in need of an *outer* criterion, or else it is a meaningless concept. We can see this even in the most ostensibly inward and subjective example of mystical experience, which also has its outer charismatic impact on others, or in supposedly 'private' qualia, which are instead the very stuff of the social communication of art.

If we are looking for the most plausible outer criteria for the very first point of emergence of a simple conscious experience in nature then this could well be in the sensitive capacity for environmental self-location as indicated by the 'outer' movement of single cell organisms. Alfred Binet (1888), Charles Sanders Peirce (1905), and the early Darwinians (see Hunt, 1995) all reached this same conclusion, that the point of emergence for simple sentience was to be found here. Binet even suggested that just as stomach cells are protozoa specialized for the function of digestion, so neurons were protozoa specialized for 'psychical attributes'. Along these lines we now know that the electrical and chemical basis of protozoan movement, potentially reflecting this immediate sensitivity to their ambient array, is identical to the depolarization along the outer membranes of neurons (Eckert *et al.*, 1988). In effect, the depolarizing flow along the outer membrane of moving protozoa would resonate, in the sense of James Gibson (1979), to the flow patterns in their surrounding medium. This capacity becomes internalized and specialized as the conductivity of central nervous systems in metazoa, linking their similarly elaborated sensory and muscular organs. Central nervous systems would not so much 'cause' consciousness as 'gather' it. As I have argued elsewhere (Hunt, 1995) this means that the focus for a science of the processes involved in the emergence of consciousness *per se* will not be within complex neurophysiology but potentially in a holistic biology of the interface of depolarization and motile sensitivity — especially so if indeterminism and spontaneity emerge independently on the level of the non linear dynamics that seem to be part of depolarization (Scott, 1995). The 'hows' of the emergence of a primary sentience would belong to a biological level of analysis, not to the neurosciences of its more complex combinations.

So there is no basis in the search for a proto-consciousness, to go still further 'inside', as with Penrose (1994), and posit the cytoskeleton of microtubules within each neuron and motile protozoan as an 'internal proto-nervous system'. Certainly protein transport along the microtubules helps to maintain the capacity for depolarization along the neuronal membrane (Hameroff, 1994), but it is equally central to all metabolic cell functions, neuronal and otherwise (Bray, 1992). Now Hameroff and Penrose make a good case for quantum effects potentially operating within and around microtubules, arguing that these will generate quantum wave effects. They even cite evidence of increased tubulin production associated with peaks of synaptic activity in new learning and its direct link to outer membrane ion channels (Dayhoff *et al.*, 1994). It is also true that neuronal microtubules are more specialized than in other cells, in that they are organized in networks in axons and dendrites. Protozoan cilia, organs of both sensitivity and movement, similarly consist in layers of microtubules (Bray, 1992). But these microtubules subserve macro-organismic functions of sensitivity and movement, as part to whole, just as, differently distributed, they support protein transport in other specialized cells. They hardly explain depolarization itself, which is more plausibly the most direct 'inner' resonance to an 'outer' sensitivity-motility. If microtubules subserve all functions of every cell, then any quantum effects demonstrable within microtubules would have more to do with basic cell coherence and metabolism, than being exclusively linked to depolarization in neurons or locomotion in protozoa. Locating the core of a primary awareness in microtubular networks is like the ancient Greeks confusing flowing blood with the phenomenal streaming of consciousness (Onians, 1951). Certainly both do flow, thus the temptation to pseudo-explanation, and certainly blood does supply the brain, but so also all other organs as well. Put otherwise, gasoline does not steer the automobile.

If consciousness is the felt 'inner' side of movement within an ambient array then quantum microtubule events go too far literally inside and are too elemental to *be* consciousness. It is not parsimonious to posit a duplicate system, even if it is temptingly 'inside' and 'hidden' in microtubules, when the more obvious inner side is already coordinated with protozoan movement and neuronal firing and may already meet the most basic criteria for simple perception. If we search for a point for the first emergence of consciousness then most parsimoniously it is still at this macro-level of organismic movement.

Epistemological Pan-experientialism

Part of the temptation of quantum models of consciousness is that they can seem *closer* to solving the 'hard problem' of 'explaining' how consciousness is possible at all, because they appear to embody some of what will emerge as its organizing principles, such as holism, complementarity and indeterminism. Whereas by contrast membrane depolarization, associated with reafferant self location in protozoans, 'merely' echoes some criteria for simple perception. But it is exactly here that the quantum approaches face their serious epistemological limitations.

Niels Bohr himself suggested that quantum effects reflect a point in the development of physics where the direct effects of observation on the phenomena observed, so akin to the limitations of our own self awareness, must blur the line between microphysics and the consciousness that would know it (Bohr, 1934). He later credited William James on the stream of consciousness as influencing both his formulation and his understanding of what he saw as the consciousness-like features of particle fields (Holton, 1968). Indeed James is strongly echoed in Bohr's 1929 discussion:

> The unavoidable influence on atomic phenomena caused by observing them corresponds to the well known change of ... psychological experiences which accompanies any direction of the attention to one of their various elements.... When

considering the contrast between the feeling of free will, which governs the psychic life, and the apparently uninterrupted causal chain of the accompanying physiological processes, the thought has, indeed, not eluded philosophers that we may be concerned here with an unvisualizable relation of complementarity (Bohr, 1934, p. 100).

James (1890) had earlier written of the way that consciousness can alternatively be seen as a collection of substantive pulses or as a continuous transitive flow, and of how attempts to isolate that flow are like trying to turn on the light quickly enough to see the darkness. Also, of course, the underlying theme of the *Principles* was the interface between the phenomenal freedom of the stream of consciousness and the determinism of nervous connections. The later James (1912) talked of consciousness and world as being two complementary ways of taking the same indistinguishable primal stuff — which on epistemological grounds he termed 'pure experience'. Others might prefer 'matter' for their primal stuff, but James, as part of the advent of phenomenology, was making the point that everything we know, including the most detailed findings of physical science, is given to us first through the medium of our experience. So whatever one's ontological or metaphysical stance (idealist or realist), epistemologically and pragmatically, experience comes first.

Paul Forman (1971; 1984) went further than Bohr in suggesting that the first discussions of the spontaneity, impalpability, and holistic properties of microphysics were actually based on explicit attempts by Germanic physicists to re-align a then much maligned 'mechanistic' classical physics with the widely accepted holistic life philosophies and phenomenologies that were attacking its reductionism. These physicists would also be aware, in their attempts to 'redeem' physics, of William James and the Wurzburg introspectionists on the impalpability and spontaneity of immediate consciousness. As Forman says, normally the fuller implications of scientific discovery in one discipline for other disciplines are only seen later and very slowly, whereas here they were simultaneous with, or even preceded, the new quantum formulations. In effect, the link of microphysics to consciousness, and the temptation, first hinted at by Bohr, to 'explain' the latter by the former, may actually rest historically on the use of phenomenology by these physicists to guide their formulations in the first place.

The point here is that it may not be legitimate to take interpretations of quantum effects that were at least in part derived from the characteristics of consciousness and then turn around and use them as purported explanations of consciousness itself. We need to be cautious in using something to explain features of consciousness that was in some sense derived from them in the first place. If the line between observer, i.e. consciousness, and observed, i.e. quantum phenomena, is intrinsically blurred, then it is not legitimate to use either one as an ontological explanation of the other — whether that is reflected in the 'idealism' of Everett's split worlds solution to the paradoxes of the quantum wave function (see Chalmers, 1996), or, facing in the opposite direction, Hameroff and Penrose's use of quantum microtubule effects to 'explain' consciousness itself. The danger in the Penrose-Hameroff approach is also a kind of projection or animism in which consciousness is first unwittingly smuggled into the properties of matter, and then these properties are used to account for features of consciousness — exactly as in the Greek ascription of the source of mind to flowing blood. Any explanatory use of the blurred line between consciousness and world is especially problematic, where physicists themselves have reached no consensus on whether the paradoxes of microphysics are ultimately more epistemological, as in Bohm's (1980) hidden variables solution, or ontological, as in Everett's literalism of infinite simultaneous universes.

Yet there is a further level of complexity here. We can still ask how it is in the first place that quantum events could be, admittedly to a degree still to be determined, *similar enough* to immediately introspected consciousness for introspectionism to actually be of help in the first formulations of microphysics. The location of consciousness-like features in quantum effects will be more than *just* a projection. After all, as Penrose (1997) has pointed out, we are not sufficiently surprised that the purely internal structures of higher mathematics can end up as suitable representations for physical phenomena not yet discovered. Similarly, as we have seen, there may be every reason to expect that some of the organizing principles of consciousness would also appear elsewhere in physical reality, given the ostensible parsimony and elegance of natural law.

David Bohm (1980) states that contrary to usual opinion, instead of modern physics being more and more distanced from ordinary experience, it actually approaches closer and closer to the principles of immediate perception. We have seen with James that immediate consciousness has its own indeterminism and complementarity, its own impalpable, unrepresentable quality. I have shown elsewhere (Hunt, 1995) that it may also be possible to locate analogues to fundamental principles of modern physics within Gibson's version of the ambient ecological array of primary perception. This might suggest that any commonalities between physics and the developed self awareness of introspective and higher meditative consciousness would reflect their mutual rooting in perception itself. For both the philosopher of science Spencer-Brown (1969) and the psychologist C.G. Jung (1959), the fact that consciousness is the necessary observational medium for microphysics means that the line between physics and psychology must blur in such a way that new developments at the limits of one can at least be *relevant* at the limits of the other. Given that the various levels of physical reality and consciousness do co-emerge in the same universe, they cannot be ultimately inconsistent with each other. We would hope, however, that any more specific coordination would be based on more than just their mutual mystery and unknowability. Yet it is precisely the epistemological limitations on our knowledge of consciousness itself that tend to be ignored in current quantum approaches to consciousness. The philosopher Martin Heidegger, lecturing in 1941, was one of the first to see this. He says, somewhat sarcastically:

> One might think physics has secured a domain for physical research in which the 'living' and the 'spiritual', and everything characterized by 'freedom', fit in perfectly ... and believe, therefore, one has penetrated into the domain of the organic. One already dreams of a 'quantum biology' grounded by 'quantum physics'.... For one is of the opinion that naturally everyone knows, off the street so to speak, what

'freedom' and 'spirit' and such things are, for one has and is these things oneself every day (Heidegger, 1993, pp. 48–9).

The deeper vulnerability then in these quantum approaches comes from the actual mystery of consciousness itself as an all-embracing medium that cannot see through or around itself and can only be selectively approximated by means of abstract metaphors taken from physical nature. We only seem to know ourselves, and later our consciousness, in the mirrored reflection of what is other than ourselves. This begins very early, in that our capacity to know ourselves by taking the role of the other is based on the initial mirroring relation between infant and mothering one, such that the infant only comes to recognize its own emotions and subjective states by seeing them reflected back in the face, tone, and gestures of its caretakers (Winnicott, 1971). Later in childhood animism, we extend this principle into physical nature and use that medium as a projective mirror for real aspects of ourselves that are outside of ordinary language. There is every indication that it is this capacity that gets internalized as the physical metaphors that cognitivists like Lakoff and Johnson (1999) see as indispensable for both abstract thought and self awareness. Certainly both the phenomenological and meditative traditions illustrate how abstract metaphors of flow, streaming, expansion, horizonal openness, spaciousness, and luminosity are needed to represent consciousness itself. I have shown elsewhere (Hunt, 1995) how mystical experience can be based on the fully felt embodiment of these same metaphors — the person going from stream of consciousness as metaphor to directly experiencing themselves as a flowing substance or as glowing light.

What all this means is that the perceptual structures we use for metaphor and symbolic representation are intrinsically bi-directional — applied outward, with suitable scientific modifications, they fit the physical universe, applied inward they both depict and *create* our experience of our own inmost nature. If it is true that we only know the features of the physical universe, blatantly so in microphysics, by discovering features of our own consciousness within it, and in turn can only represent our consciousness with metaphors from a physical universe so uncovered, then there will be epistemological limits in both directions. If so, it is a mistake for microphysics or consciousness studies to confer ontological explanatory status on the features of the other medium. Neither traditional idealism nor traditional realism will do here. The intersection and blurring of microphysics and phenomenology is not the place of cause and effect, but rather of the relation between lens and referent in metaphor theory. This relation potentially goes in both directions, consciousness a metaphor for quantum effects and microphysics a lens to illuminate, but not explain, consciousness. There is still plenty of room to locate similar principles of organization on different hierarchic levels of the universe, including consciousness, without that making any one of them the direct cause of the others — any more than the spiral patterns of galaxies 'cause' the similar spirals in the shells of snails or vice versa. The ways that phenomenology and physics can inform each other epistemologically need to be kept separate, despite the difficulty, from ontological claims of explanation.

Some Concluding Reflections

Perhaps we can paraphrase Ernst Mach (1959) and say that there are ultimately only two disciplines with claims on an all-embracing perspective: physics and phenomenology. Yet in the very midst of their claims of ontological primacy, each finds itself with a curious dependence on the other. William James (1912) suggested that the extremes of objectivity and subjectivity are ultimately based on contrasting attitudes we can take towards a shared or common stuff — a primary beingness which he saw as a sheer 'thatness' not yet taken as any sort of specific 'whatness', whether as the 'whats' of empirically describable consciousness or of the physical universe. James, as psychologist and phenomenologist, preferred to call the primary thatness 'pure experience', reflecting his pragmatic stance that all knowledge comes through the lens and medium of our consciousness. Physics of course understands this 'stuff' of everything as ultimately based on quantum effects.

With respect to the sciences, consciousness, as a natural system of maximum hierarchic complexity, relies on and re-uses multiple less complex physical systems, including quantum reality, as well as molecular chemistry, as part of its higher order emergence. As such we could say that consciousness, as system, is the maximally complex, knowable expression, and so 'point', of the universe — within which it must have *in some sense* been latent or potential from the beginning (see also Jonas, 1996). While this notion of the consciousness of living organisms as already latent within the 'big bang' might seem a mere truism, it may be nonetheless a better guide to widely accepted notions of emergence in modern science than the older model of a mechanistic 'chance' somehow ending in such a complex hierarchic system as human experience. So it is of interest that phenomenology, when pushed to its furthest development in the meditative traditions, leads to direct experiences in which the universe is felt to be self-aware, and consciousness in turn is felt to be the very point and purpose of Being itself (Almaas, 1996). Again we see that the maximum development of physical systems theory and experiential phenomenology approach each other in an *intuitive* unity. I have suggested that this is best *understood* in Jamesian pragmatic terms rather than via the unprovable idealist ontology of Whitehead or the reductionist physical ontologies of quantum explanations of consciousness.

Addendum

Critics of this view — that consciousness, once emerged, is *sui generis* and not explicable in terms of its parts, and therefore constitutes a conceptual 'primitive' — dismiss this perspective as 'mysterian'. Apparently they forget that space–time itself similarly emerges at an 'earlier' point out of an equally mysterious singularity. If space–time is a categorical 'primitive' that once emerged can only be studied in terms of its variations and organizing principles, then consciousness, as a system emerging later in the unfolding of the universe, may still have a similar ontological irreducibility (see also McGinn, 1995). Certainly experience has an epistemological primacy as a medium, and that every bit as functionally all encompassing for self knowledge as space–time seems to be for the material universe. Once so emerged, a reflective consciousness will have no choice but to find the latent seed of itself dispersed through less complex physical systems — a key and legitimate point in Whitehead's (1929) philosophy of science. Where the present approach differs, however, from ontological panexperientialism is that the latter confuses these dispersed or 'submerged' seeds and principles of a later emerged consciousness with consciousness itself.

There is no logical reason, given contemporary notions of emergence in science (Sperry, 1987; Scott, 1995), why an evolving universe would have to produce all its categorical primitives 'at once'. So far, and by definition at the very limits of our understanding, we would have two such fundaments, as revealed in two questions that are inherently unanswerable from the perspective of science, and so showing its limits: 'Why is there something rather than nothing?' and 'Why is there a consciousness of that something?' Science is in the business of tracing the variations and underlying principles of matter and mind, but it cannot give explanatory answers to *those* questions without becoming theology. A puzzle here, however, is why our era would think that the basic categories of Being would have to 'appear' simultaneously. Isn't there Becoming too? Thus we are lead to the emergentist ontology of a *sui generis* consciousness and the conjoined epistemological or pragmatic pan-experientialism advocated herein as the most balanced way of threading through current debates over the nature of consciousness. The so-called 'hard problem' of 'explaining' how something material can 'cause' consciousness may be closer to an unwitting theology, and certainly an unconscious and unprovable metaphysics, than its explicators have realized. As Wittgenstein said, 'explanations come to an end' — certainly for physicists and maybe even for psychologists.

References

Almaas, A.H. (1996), *The Point of Existence: Transformations of Narcissism in Self Realization* (Berkeley, CA: Diamond Books).
Binet, A. (1888), *The Psychic Life of Micro-organisms* (Philadelphia: Albert Saifer, 1970).
Bohm, D. (1980), *Wholeness and the Implicate Order* (London: Routledge and Kegan Paul).
Bohr, N. (1934), *Atomic Theory and the Description of Nature* (Cambridge: Cambridge University Press).
Bray, D. (1992), *Cell Movements* (New York: Garland Publishing).
Chalmers, D. (1996), *The Conscious Mind* (Oxford: Oxford University Press).
Dayhoff, J., Hameroff, S., Lahoz-Beltra, R. & Swenberg, C. (1994), 'Cytoskeletal involvement in neuronal learning', *European Biophysics Journal*, **23**, pp. 79–93.
Eckert, R., Randall, D. & Augustine, G. (1988), *Animal Physiology: Mechanisms and Adaptations* (New York: Freeman).
Forman, P. (1971), 'Weimar culture, causality, and quantum theory, 1918–1927: Adaptation by German physicists and mathematicians to a hostile intellectual environment', in *Historical Studies in the Physical Sciences*, ed. R. McCormmach (Philadelphia: University of Pennsylvania Press).
Forman, P. (1984), '*Kausalität*, *Anschaulichkeit*, and *Individualität*, or how cultural values prescribed the character and the lessons ascribed to quantum mechanics', in *Society and Knowledge*, ed. N. Stehr & V. Meja (New Brunswick, NJ: Transaction Books).
Gibson, J.J. (1979), *The Ecological Approach to Visual Perception* (Boston, MA: Houghton Mifflin).
Hameroff, S. (1994), 'Quantum coherence in microtubules: A neural basis for emergent consciousness?', *Journal of Consciousness Studies*, **1** (1), pp. 91–118.

Hawking, S.C. (1988), *A Brief History of Time* (Toronto: Bantam Books).
Heidegger, M. (1993), *Basic Concepts* (Bloomington: Indiana University Press).
Holton, G. (1968), 'The roots of complementarity', *Eranos Jahrbuch*, **38**, pp. 45–90.
Hunt, H. (1995), *On the Nature of Consciousness: Cognitive, Phenomenological, and Transpersonal Perspectives* (New Haven, NJ: Yale University Press).
James, W. (1890), *The Principles of Psychology* (New York: Dover).
James, W. (1912), *Essays in Radical Empiricism and a Pluralistic Universe* (New York: Dutton).
Jibu, M. & Yasue, K. (1997), 'Magic without magic: Meaning of quantum brain dynamics', *The Journal of Mind and Behavior*, **18**, pp. 205–27.
Jonas, H. (1996), *Mortality and Morality: A Search for the Good after Auschwitz* (Evanstan, IL: Northwestern University Press).
Jung, C.G. (1959), *Aion: Researches into the Phenomenology of the Self* (New York: Panetheon Books).
King, C. (1997), 'Quantum mechanics, chaos, and the conscious brain', *The Journal of Mind and Behavior*, **18**, pp. 155–70.
Lakoff, G. & Johnson, M. (1999), *Philosophy in the Flesh* (New York: Basic Books).
Mach, E. (1959), *The Analysis of Sensations* (New York: Dover).
Miller, G. (1981), 'Trends and debates in cognitive psychology', *Cognition*, **10**, pp. 215–25.
McGinn, C. (1995), 'Consciousness and space', *Journal of Consciousness Studies*, **2** (3), pp. 220–30.
Natsoulas, T. (1983), 'Concepts of consciousness', *The Journal of Mind and Behavior*, **4**, pp. 13–59.
Onians, R.B. (1951), *The Origins of European Thought about the Body, the Mind, the Soul, the World, Time, and Fate* (Cambridge: Cambridge University Press).
Peirce, C.S. (1905), 'The principles of phenomenology', in *Philosophical Writings of Peirce*, ed. J. Buchler (New York: Dover, 1955).
Penrose, R. (1994), *Shadows of the Mind: A Search for the Missing Science of Consciousness* (Oxford: Oxford University Press).
Penrose, R. (1997), *The Large, the Small, and the Human Mind* (Cambridge: Cambridge University Press).
Rosenblum, B. & Kuttner, F. (1999), 'Consciousness and quantum mechanics: The connection and analogies', *The Journal of Mind and Behavior*, **20**, 229–56.
Scott, A. (1995), *Stairway to the Mind: The Controversial New Science of Consciousness* (New York: Springer-Verlag).
Sheets-Johnstone, M. (1998), 'Consciousness: A natural history', *Journal of Consciousness Studies*, **5** (3), pp. 260–94.
Spencer-Brown, G. (1969), *Laws of Form* (New York: Dutton).
Sperry, R.W. (1987), 'Structure and significance of the consciousness revolution', *The Journal of Mind and Behavior*, **8**, pp. 37–66.
Whitehead, A.N. (1929), *Process and Reality* (New York: Macmillan).
Winnicott, D.W. (1971), *Playing and Reality* (New York: Basic Books).
Wittgenstein, L. (1980), *Remarks on the Philosophy of Psychology*, vol. *1* (Oxford: Basil Blackwell).
Wittgenstein, L. (1992), *Last Writings on the Philosophy of Psychology*, vol. *2* (Oxford: Basil Blackwell).

Natika Newton

Emergence and the Uniqueness of Consciousness

This paper argues that phenomenal consciousness arises from the forced blending of components that are incompatible, or even logically contradictory, when combined by direct methods available to the subject; and that it is, as a result, analytically, ostensively and comparatively indefinable. First, I examine a variety of cases in which unpredictable novelties arise from the forced merging of contradictory elements, or at least elements that are unable in human experience to co-occur. The point is to show that the uniqueness of consciousness is comprehensible in terms of a more general kind of emergence. I then argue that phenomenal consciousness essentially involves synchronous activations of representations of 'identical' intentional objects with distinct temporal tags, and is thus a case of the emergence of novelty from forced blending of incompatible components. It follows from the general nature of such emergence that consciousness would be indefinable and hence seem mysterious. This analysis will show why phenomenal consciousness would be impossible to resolve into its constituents by the conscious subject. The result is, I hope, a happy blend of physicalist explanation with respectful acknowledgement of the robustness of subjective experience.

I: Introduction

This paper looks not at phenomenal consciousness itself as a unique mental entity, but instead at human ways of making sense out of experience for purposes of integration into pre-existing goal structures. Most experiences are made sense of in relation to other types of experience. They are defined by analysis into their experiential components (e.g. feelings of anxiety into bodily sensations and imagery); by ostension (as when the experience is of a public object that can be pointed to); or by comparison (as with colour, which is a particular type of visual experience, each colour occupying a position along a familiar spectrum). Any experience immune to all of these will be a mystery to its subject. There is only one experience of which that is completely true: phenomenal consciousness. All the above examples are of *types* of phenomena, understandable as one among others of that type. But phenomenal consciousness itself is *sui generis*. Nothing else is like it *in any way at all*, because anything other than phenomenal consciousness is unconscious, and hence not like anything.

Understanding why phenomenal consciousness is unique for subjects will obviate the need to find a special ontological or epistemological explanation for it. Instead, we will understand why there is nothing of the kind to be found, and why the apparent uniqueness of the phenomenon is simply the predictable result of certain brain mechanisms.[1] At present, the only way to categorize consciousness as one of many in any category is generically, as an emergent property. A property is emergent if it holds of its constituents collectively, but not individually, and if it comes into existence only through the combined actions of the constituents. But many aspects of novel emergent properties are unpredictable from the properties of their constituents. Some, like temperature, can be explained in a way that makes it clear exactly how they emerge, and hence can be completely demystified. But emergent properties are not always so tractable. So far, that has been the case with phenomenal consciousness. One result is a tendency to view the emergent property of consciousness as nonphysical.

The position of this paper is that the particular ineffability of phenomenal consciousness in general (irrespective of its content, which may be perceptual, introspective, or something other) results from its unique indefinability. This is a semantic problem, not an ontological or even an epistemological one. As such, the problem provides no reason to view consciousness itself as intrinsically of a different nature from other physical emergent properties.[2] The indefinability of consciousness is in part a result of its subjectivity. This term, in turn, needs to be demystified. I define a phenomenon as subjective if it is a response to an organism's own proprioceptively-sensed self-generated activity, rather than to publically-recognized 'external objects'. Subjectivity on this account is no more than limited access of a particular physical kind, and not a nonphysical form of experience. It is, moreover, not the same thing as consciousness: subjective events as here defined may be conscious or not.

The above definition of subjectivity is not commonly formulated as it is here, but it is not radically novel. It is entailed by any theory holding that a subject's experience is produced by sensory mechanisms that are in principle accessible by another subject. It has been traditionally held that subjective experiences are those available only to a single subject. But if subjective experiences are experiences of states of the subject's body, which is not at all a radical claim, then their subjectivity, or privacy, is an accident of neural wiring. As I noted in an earlier paper on pain (Newton, 1989),

---

[1] Dennett and other eliminativists draw conclusions that might be justified if they made this approach explicit; since they do not, they appear not to recognize the rich experiential nature of phenomenal consciousness. It will not do to say: there is no such thing as a quale. One must first explain why there can appear to be such a thing, and for that one must characterize the experience in an intuitively acceptable way.

[2] To say that the problem is semantic rather than ontological or epistemological does not in itself entail physicalism.

The privacy that we must allow to pain does not threaten pain's status as a secondary quality, since it is simply the logical consequence of the pain's being a property of one's own body. Importantly, this sort of privacy does not entail phenomenal properties of the pain accessible only to its possessor.... [W]hat my pain is like is what your pain is like if the two instances of pain result from your nerve endings implanted, together with my nerve endings of the same type, in a single shared body part (Newton, 1989, p. 577).

Phenomenal consciousness is a species of subjective novelty, which occurs when an organism, by means of self-motivated acts of attention, 'blends' distinct but incompatible (in some sense) sensational components — e.g. features of the perceptual activity, such as location in the visual field, rather than of the perceptual object — of perceptual acts into a unified representation, the 'intentional object' of experience.[3] Blending in this sense is the experiential correlate of neural binding through synchronized firing or re-entrant loops (Edelman, 1989; Crick and Koch, 1990; Damasio, 2000). The resulting blend is best viewed as a particular type of emergent property. Not all binding results in subjective novelty, of course; my drearily familiar perceptual experience of the computer I am now using is the result of binding. But binding can also produce unified experiences whose components, when considered independently, seem fundamentally discordant. Synaesthesia is a striking example (Cytowic, 1993). If we consider phenomenal consciousness in general (as distinct from conscious experiences in particular modalities) in relation to other types of emergent property, we may understand what gives it the uniqueness described above: it cannot be defined analytically, ostensively, or comparatively. Understanding that will, in turn, demystify it.

The term 'unified experience' deserves a caveat. It is almost universally taken for granted that conscious experience has an actual, objective unity, although perhaps apparent only to the subject. I consider that assumption ambiguous and questionable and do not make it here. What is clear is that under normal conditions we experience our synchronous conscious states *as unified*, or 'co-conscious' rather than fragmented. I shall not assume that the explanation for this experience is some conceptual unification of the components of conscious states themselves (although the underlying brain mechanisms may be bound together through such events as synchrony and reentrance), but that it could instead lie entirely in a superficial, unanalysed and confused impression or assumption on the part of the subject. A more detailed account of this hypothesis is provided further on.

In self-organizing creatures, internal states can embody inherent contradictions or incompatibilities, either in absolute logical terms or relative to the structure of the organism. Sometimes these contradictions are untenable, and one of the states must give way — e.g. the motivation to move in opposite directions at the same time. In other cases, however, the organism can or must incorporate both states

---

[3] A question arises as to whether involuntary acts of attention fall under this definition. The position here, which is compatible with much recent research, is that while the capturing of attention may occur outside of consciousness, deliberate prolonging of attention is normally a conscious act.

into a higher level of organization. There are many ways in which we describe states of an organism as incompatible. They can be fundamentally physically incompatible, as in the above examples. We may describe them as neurologically incompatible, as when activating and inhibiting neural activations compete for dominance of a particular brain system. Then, in sensate organisms we may see them as perceptually or experientially incompatible, e.g. when one must focus on figure or ground, but not both. Finally, in beings capable of cognition, they can be incompatible in their intentional content. The man seen in the distance might be the dean or the president, but not both. In systems of logic, incompatibility is logical contradiction: within a specific context a statement cannot be both true and false. There are various ways these incompatibilities are resolved, a full discussion of which is impossible in this paper. The one important here is the way we humans *can*, both consciously and unconsciously, incorporate incompatibilities of potential intentional content into novel unities. As we often say, we can 'agree to disagree'.

The point can be made in terms of the notion of 'spaces' (Fauconnier, 1985) in cognitive linguistics. The incompatibility at issue here lies in their original inclusion in a space whose structure does not allow for their co-existence, or unity. They can therefore be united only in a distinct, more inclusive space. This view can be applied to logical contradiction. A statement cannot be both affirmed and denied within the same context, or space; but one can adopt a wider context in order to answer the question 'Did you enjoy the party?' by saying 'Yes and no.' The wider context (one goes on to explain) is that of one's total state, encompassing both one's physical pleasures and one's emotional discomforts (one might say 'I had mixed feelings'). It is true that the word 'enjoy' is itself ambiguous, and so might be thought to accommodate both contexts at once. But the question doesn't have a simple yes-or-no answer unless both contexts give the term the same value, and often they do not.

I shall argue that phenomenal consciousness is the result of the incorporation, into a single representational schema, of representations of organismic states that are incompatible because the states are represented as occurring at different times. Two such states, for example, are those of expecting to perceive x and actually perceiving y. My actually seeing Mary walk through the door at time t is incompatible with my expecting — believing that I will see — John walk through the door at t. The former refutes the latter: if I had known that Mary would walk through the door, I would not have expected to see John. It is a surprise to see Mary precisely because my expectation of seeing John is still consciously active; seeing Mary and expecting to see John (perhaps by means of a vivid visual image of John) are blended in conscious experience. The conscious state of surprise, we might say, is an emergent property of that blend.

Most of our conscious states are not states of surprise, but otherwise the above description seems to apply. A sense of self-continuity, which allows intelligent action, requires awareness of some of my past to be concurrent with awareness of new stimuli. In other words, when for purposes of goal selection and planning I attend to my current situation, I do so by blending memories of temporally distinct states into a single, extended 'present'. The unification in a single representation of 'contradictory' states ('now' and 'not-now') produces a novel state with the experiential properties of phenomenal consciousness.

Briefly, therefore, my conclusion is twofold: first, that consciousness, like many emergent properties, arises from the forced blending of intrinsically incompatible components into a distinct unified framework; and second, that consciousness is a unique type of emergent property that is analytically, ostensively and comparatively indefinable. The plan of the paper is as follows. First we look at the general issue of incompatibility and novelty. In line with the goal of categorizing consciousness as one among other types of emergent property, we examine a variety of cases in which unpredictable novelties arise from the forced merging of incompatible elements. The point is to show that while consciousness may be unique as a psychological state, its uniqueness is comprehensible in terms of a more general kind of emergence. I then argue that phenomenal consciousness essentially involves synchronous activations of representations, with distinct temporal tags, of more or less 'identical' intentional content. The greater the identity, or matching of expectations with actuality, the less the surprise or confusion. Understanding this brain process allows the prediction of certain experiential properties of phenomenal consciousness that identify it as a case of the emergence of novelty from incompatibility. I argue that it follows from the general nature of such emergence that the properties of consciousness would be indefinable and hence seem mysterious, and even incoherent, which in fact they are. While this analysis will not result in the elimination of those properties from subjective experience, it will make clear why phenomenal consciousness would necessarily be impossible to resolve into its constituents by the subject. The result of this discussion is, I hope, a happy blend of physicalist explanation with respectful acknowledgement of the robustness of subjective experience.

II: Incompatibility and Emergence

For present purposes emergent phenomena can be divided into two kinds: collective behaviour, such as fluidity, that is part of a scientific description of the physical world; and collective properties, such as wetness, that expressly invoke the sensory experiences of observers. In both cases descriptions of the emergent phenomena require new vocabularies not applicable to the constituents of the phenomena. In the former, however, the novel phenomenon is completely definable in quantitative terms (e.g. fluidity is the capacity to move or change shape without separating when under pressure). In the latter, a description of the phenomenon requires novel qualitative terminology; a complete description essentially includes reference to unique sensations produced in observers. Since the sensations are qualitatively unique, the description is not analysable into descriptions of component sensations.

Not all emergent phenomena arise from contradictions or incompatibilities among their constituents. Some, like fluidity and other collective physical behaviours, come into being naturally through lawful, coherent interactions of physical objects. But others are forced: when a single framework *must* include intrinsically incompatible parts, then there is an inevitable tension in the new unity. There is not room here for a full discussion of these various conditions of emergence. Our focus is the framework that constitutes phenomenal consciousness, and the inevitable tension — the sense of mystery, insubstantiality, or ineffability — that it embodies. We emotionally-driven self-organizing beings are the agents of this particular kind of forced blending; but there are other examples. In the following section I discuss several cases of novelties that emerge from incompatibilities. I start with a conceptual novelty as a purely formal example to highlight the general point; I then consider cases involving phenomenal experience.

Russell showed that Frege's set theory would require the existence of sets that both are and are not members of themselves, and hence was inconsistent. In this case, the inconsistency lies in its incompatibility with our usual rules for thinking. Various methods for resolving the paradox have been proposed. Russell's solution involved a cumulative hierarchy of types (Russell, 1908), while others rely on the Axiom of Foundation. In each case the effect is to prohibit the membership of a set in itself. While not unique, each of these solutions may be considered to have emerged from the paradox.[4]

A different example is physiological: the anti-contrast 'assimilation effect' in the perception of coloured areas (Hardin 2000, p. 115). In retinal based colour perception, when coloured areas are very small, their colours are blended: 'Thus, although a large region of red is sharply accentuated when placed next to a large area of green, when the areas are tiny, the retinal receptors blend the red and the green to give the appearance of a pure yellow.' In this example, I take the appearance of yellow to be an emergent property whose phenomenal nature — that is, what it is like to see yellow — would not be predictable by the subject from previous experiences of unblended red and green. Of course, the usefulness of this example is limited for two reasons. First, the experience of yellow normally results from the positive value of the activation of the post-retinal yellow/blue channel along with the equilibrium of the red/green channel. Thus I cannot claim that this effect of seeing small areas of red and green would yield the appearance of yellow in the absence of the already-existing yellow/blue channel; it may depend on one having already experienced yellow. Second, a more serious limitation: all individual colours, as consciously experienced, are agreed to be ineffable and ultimately mysterious, so in that sense yellow is no more so than any other. That is a problem with most examples that could illustrate my point: each of them is both an example of conscious experience *per se*, and also an example of a particular category of conscious experience having its unique phenomenal mystery in addition to that adhering to all phenomena. When, as here, I am referring to the narrower uniqueness, I invite confusion of that with the more general. I want to include this example, nevertheless, because I want to portray the variety of kinds of emergence. The example serves as an illustration by way of a very narrow

---

[4] I am particularly indebted to Joseph Goguen for helping me with the preceding discussions of incompatibility.

analogy: as a result of the potential activation of two incompatible brain states, a mechanism produces in the subject an experience associated with a state that is qualitatively distinct and unpredictable from what would be produced by either of the incompatible parts. In this particular case, of course, the experience of yellow occurs *because* the subject cannot be independently conscious of the assimilated colours. (It should be emphasized that [conscious] colour perception is just another kind of conscious experience, not a different example of emergence. As I explain further on, phenomenal consciousness can vary in its content, and different elements — from proprioception as well as exteroception — can capture the subject's focus at different times.)

The above examples share this feature: two states (logical constructions or biological processes) that are in important ways incompatible are maintained in juxtaposition. If they occurred only serially, the emergent solution would be unnecessary. We can alternate between the perceptions of two incompatible perspectives, as in the Necker cube. The emergent property requires forced simultaneous processing of two incompatible states or representations, long enough for the organism to register the incompatibility and attempt to resolve it; the emergent property is the result.

Consider examples more purely experiential. Depth perception is an obvious one (Engel *et al.*, 1999). Binocular vision presents two visual fields, isomorphic but slightly displaced. The result is not a blur or a confusing double image, but visual depth (this analogy was first suggested in Newton, 1991, p. 34; see also Newton, 1996, p. 186). It is not deducible from binocular processing alone that visual depth would occur; we can imagine instead the experience of Necker cube-like aspect shifts. Visual depth emerged, we may assume, for adaptive reasons. The same is true of motion perception: areas of the inferior temporal lobe blend successive visual sensations into representations of a single moving object. *Ex post facto*, motion perception makes sense as necessary for our type of organism, but its phenomenal properties are not predictable from binocular vision considered in isolation.[5]

I will look at one more emergent phenomenon before discussing phenomenal consciousness itself. Our experience (and concept) of persons is such a phenomenon, and like the others appears to result from the forced blending of incompatibilities. Cells in the parietal and motor areas are believed to respond to two incompatible or mutually exclusive conditions: the experience of deliberate bodily movement of the subject, and the sight of the same kind of movement by a conspecific. The result is the representation in a single structure of both inside (subjective) and outside (objective) aspects of behaviour — what it feels like and what it looks like — a single being as both an agent and an object of agency (a visible object) (Barresi and Moore, 1996). Our concept of a person is a concept of such a being, and many now think it plausible that its semantic ground lies in the activity of these 'mirror neurons' (Gallese *et al.*, 1996). We are able to have the

---

[5] There is this exception: if one already knows in advance from one's own case what a qualitative property is like, and if one knows the brain states with which it is correlated, then observing those brain states in another person would allow one to predict, with varying accuracy, the qualitative properties.

concept because we are able to generate that unique experience. Note that the emergent experiential concept of a person as having both objective and subjective aspects is not itself objectively unifiable: objective and subjective (outer and inner) perspectives on the same entity are mutually exclusive, at least within our common conceptual framework. Nevertheless, we do experience 'personhood' in a unified way, at least when we do not try to analyse the experience.

III: Phenomenal Consciousness and Multitemporal Representation — The Extended Present

It is widely held that phenomenal consciousness is correlated with synchronous firing of neurons and reentrant signalling in several cortical areas (e.g. Tononi and Edelman, 1998). These mechanisms unify the activities in those areas, such that human subjects report corresponding experiential awareness. If the organism is in a particular state of sensory activation, for example, the subject will report that she sees an object; in normal vision the perceived object is correlated with the external stimulus to which her visual system has attuned itself. The newly conscious state, in turn, allows the organism more degrees of freedom to select future actions, since its own responses can now be represented as explicit goals subject to rational planning, rather than remaining behind-the-scenes approach/avoidance motivators. What is it about synchronized neural activity that makes such a difference to the subject? The brain change is quantitative only, but for the subject the change is a qualitative leap — in fact, the change introduces quality for the subject, where none existed previously. My proposal is that the experiential aspect of phenomenal consciousness is one result of the unification of representations, in the way we actually do this, of distinct temporal moments into a single representational framework, which as a result gains what we call 'temporal thickness'. The unified representation is experienced as 'now' by the subject: as the present moment comprising her own immediate interaction with the environment.

Note that I do not claim that temporal thickness is all there is to consciousness, or that understanding it elucidates all of the traditional puzzles. I am claiming only that temporal thickness constitutes its unique and most intransigent aspect. That aspect is the 'what it is like' phenomenon — the property of all conscious states to be 'like' anything at all, to be phenomena. Some might object that phenomenal consciousnes includes more than just temporal thickness, such as unity, lack of granularity in visual field, etc. But while it does include those, they wouldn't *be* phenomenally conscious without temporal thickness, even if the mechanisms producing them were working: temporal thickness is what allows all these other things to last long enough to be conscious. The 'more' that consciousness includes are emotional drives and other stimuli to the planning of action, and all the purely sensational aspects of conscious perception, topped off by the intentional objects we are 'conscious of'. But the ability of all those things *to be phenomena*, once they're in place (if that can be isolated at least conceptually from temporal thickness) is the result of their being extended in time. This aspect is what is referred to as the 'hard problem' (Chalmers, 1995). Other puzzles, such as spatial awareness, dream consciousness, possible divided consciousness in split-brain patients, etc., remain.

This idea is closely related to that of Nicholas Humphrey (1992), who theorizes that the construction of an 'extended present' from feed-back loops underlies phenomenal consciousness. Our continual sensorimotor exploration of our environment has the result that

> what constitutes the conscious present is largely the immediate sensory *afterglow* of stimuli that have just passed by — the dying-away activity in reverberating sensory loops. And it would follow that the temporal depth and subjective richness of this conscious present is bound to be determined by just how long this activity survives (Humphrey, 1992, p. 189).

Humphrey goes on to diagram quite elegantly the differences among various degrees of consciousness as they correspond to various durations of the reverberating sensory activity, from mescaline-induced 'consciousness expansion' to the ultimate shrinkage of the present in sleep.

Thus the extended present has a completely nonmysterious physiology. Experiencing 'now' entails a contradiction, however. As James also famously noted, at each moment of experience, the 'now' of which we are aware is in fact the 'just past', the previous moment in which we were anticipating and preparing for 'now'. In that previous moment, we activated anticipatory imagery of the actual present moment, and this anticipatory imagery is experienced again now, as anticipatory, along with the sensations that were anticipated. Novel sensations, in other words, are experienced *as having been anticipated*. It is not simply that we remember anticipating them; rather, their anticipation is a part of them as currently experienced, which accounts for the feeling that what we are now experiencing has been there 'already'.

To be sure, novelty by its nature is not anticipated, if by that one means that before it arrives, one expects it to appear exactly as it does appear. But if there was no preparation at all for novel stimuli then, as Mack and Rock (1998) have shown, there is no attention to it. Normally when one notices something unexpected, one has been expecting something else, and his thus *aware of the novelty as different* from what was expected. One might have even been expecting nothing at all, and is hence 'ready' for silence and emptiness. Then the novel stimulus both does and does not satisfy the expectation, and that means both that something was expected, and that the actual novelty can be appreciated for its unexpected uniqueness, since attention has already been drawn to its possibility.

That phenomenon in which the novel stimulus is experienced as having been anticipated, but not in its actual form, is a result of the blending of distinct temporal moments by synchronized neuronal firing. Because our present experience includes (at least) two distinct times, it is experienced not as an instantaneous slice of time but as an extended time, containing elements of both 'now' and 'not-now', in a unified immediate representation. This unification allows the novelty of the stimulus to be fully appreciated, since the prior state of inaccurate anticipation lingers along with the new input. The entire series is still experienced, however, as 'the present'. It has been proposed that the binding via synchronized oscillations is essential for phenomenal consciousness (Crick and Koch, 1990). It is not obvious, however, that such binding *per se* can explain the uniqueness of consciousness, without the recognition that the unification of distinct times in one present moment is a contradiction for the subject, who retains a sense (and, usually, a concept) of other past times as past. 'Past' and 'present', understood conceptually, remain distinct in experience — the contents of both working memory and long-term memory retain their character as no longer present. But *'the present' as experienced* includes both some past times of its own, as well as anticipations of the future (Ellis and Newton, 1998, p. 425). The past that is part of the subject's experienced present is not identified by the subject as moments that are past and gone. Instead, it feels like a part of the present, one that gives the present a significance and a substantiality that it could not have if it consisted only of an instantaneous time-slice. And in the same way, the anticipated future is also there in the present moment, allowing us in that moment to anticipate its continuation and to formulate intentions, as for shifts of attention or bodily interactions with the currently-perceived objects, to take place in this temporal space we want to call 'now.' Thus as we experience it, the present moment is a logical impossibility, a chimera. Yet it constitutes reality for the subject. As Descartes observed, it is that of which we can be most certain.

I will describe an artificial slow-motion or frame-by-frame version of what I have in mind (the process would in fact be a dynamic one, not neatly divisible into distinct stages). Imagine that you have just awoken, and your senses respond to stimuli from your environment. You are too confused to identify where you are or what is happening, but your brain may be constructing, at time t_1, a representation composed of visual data, somatic data, efference copy, emotion, and other elements of your current state. Then, at a later time t_2, your brain constructs a second representation, which contains all of the same elements, updated, PLUS the representation from t_1, which is retained in working memory. What do I mean by 'PLUS'? I will venture this: the t_2 representation is an act of attention to only those elements of information that resonate with the t_1 representation. Now suppose we say that the two representations together constitute a conscious state. (We must assume that all other necessary supporting brain mechanisms are in place.) The act of attention binds the two representations into a single 'object': the present moment. This single 'object' is formed from two components that, by their physiological nature, should not co-exist: events at both t_1 and t_2. But they are posited as coexisting, in that their combined representation is labelled as occurring only at t_2. That means that the representation at t_2 is the experience of looking at, as it were, not just remembering, the past. Or, it is the experience of the present as extending into the past.

The present, in other words, is experienced as temporally thick. This thickness varies. In the above example, only two 'moments' were described for purposes of illustration; in fact much more of the past is with us always. The experience of this temporal thickness is the emergent qualitative property that is phenomenal consciousness. Because it is qualitatively distinct from other emergent properties, it cannot be defined analytically, in terms of any qualitative components. It is, moreover, subjective, in the sense that it arises from an act of attention to features of the agent's own perceptual activities, rather than to public features of perceptual objects. Because it is subjective and not public, it cannot be defined ostensibly. And because it is unique in applying to all and only conscious experience, rather than being one phenomenal property among others, like colour or taste, it cannot be defined comparatively. There is nothing else relevantly similar to phenomenal consciousness. It follows that phenomenal consciousness would necessarily feel like a mysterious and ineffable property; and it also would follow that people unversed in brain mechanisms would be at a loss to explain it in terms of any other property. It does not follow, however, that consciousness cannot be explained fully in terms of brain mechanisms.

Throughout the above discussion, I have been ignoring two important dimensions of conscious experience: there are degrees of consciousness (a quantitative dimension) and also variations in the focus of consciousness (a qualitative dimension). Sometimes one's conscious awareness seems rich, intense, and long-lasting; at other times one is conscious only fleetingly. This variation may be a function of both the duration and the scope (how much remembered material is included in the blend) of the conscious state. One's focus can vary during the same continuous state of conscious; when we introspect, for example, our focus shifts from external objects to the private states of our bodies, and this shift can occur during a single extended period of consciousness. It may be that in so-called 'peak experiences', one's focus is extremely narrow and excludes proprioceptive sensations. It is important to note that in speaking of the 'uniqueness' of consciousness in this paper, I am not referring to self-conscious alone, but to what all states of consciousness have in common: the 'what-it-is-like' feature. While bodily awareness seems to be a background element in all conscious experience (Damasio, 2000; Panksepp, 1998), it can and frequently does fall outside the focus of attention. In those cases we may be said to be conscious 'only' of external sensory input. But while we may speak in that way, the very notion of *externality* presupposes some awareness of one's own body and its boundaries, as does any awareness of possible interactions with the external world — its 'affordances' (Gibson, 1986). This issue is treated in more detail in Newton (1991).

IV: Conclusion

Throughout this discussion I have spoken of the 'subject' of experiences, and of representations, in a way compatible with an 'act-object' view of mental experience: that subjective experience is the mental act of 'perceiving' a mental object. This view entails that there is a kind of parallel universe to the ordinary physical one. In the ordinary one, physical bodies (including sense organs) act on (e.g. see) external physical objects. In the mental one, there are mental acts (thoughts) performed on mental objects (qualia, ideas), that are accessible only to the subject. There are, however, fatal objections to such a view (see Newton, 1989; Ellis and

Newton, 1998) and there is an alternative that is also compatible with the present proposal. The primary objection to an 'act-object' view of subjectivity is the threatened regress, as well as the rhetorical question: *who or what* is the subject of conscious experiences, if that entity is distinct from the experiences? An important goal of this paper has been to provide an alternative to the act-object metaphor, one drastic enough to destroy the tendency of that metaphor to come creeping back in to theories of consciousness.

In the usual syntax in which 'phenomenal experience' or 'phenomenal consciousness' is used, they are direct objects, but this reading is completely unnecessary. We can say instead that states of neural activation become conscious states under particular conditions of synchronization and reentry. To be a conscious state does not entail having an observer. It can instead simply entail a form of existence such that the organism in which it obtains undergoes an especially comprehensive form of integration, one that allows reflection and planning that are impossible in states of lesser coordination. We do say that being conscious feels a certain way 'to me', but such expressions cannot bear much metaphysical weight. What feels certain ways 'to me', literally, are the objects of the individual sensory states that are the components of states of phenomenal consciousness (see the admirable essay by Austen Clark, 1998).

Substituting what we might call a 'self-maintaining' view of conscious experience for an act-object view (in which the 'object', phenomenal experience, requires the act of attention by a subject in order to be conscious) supports the position of this paper that conscious experience is an unstable and temporary synchronized activation of various intrinsically incompatible components. One of its components may well be the sensation of visual access to external objects, and this component may imbue the whole experience with an illusory act-object quality (Newton, 1991). It is perfectly coherent to speak of a conscious mental state possessing illusory properties without committing oneself to the existence of a separate subject experiencing that state: we need say only that the state functions within the cognitive system of the organism as though it is conceptually coherent, when in fact it is not. Allowing such a possibility will allow us to slice through a tangle of philosophical questions about how such a mysterious entity could possibly be physically realized.

References

Barresi, J. and Moore, C. (1996), 'Intentional relations and social understanding', *Behavioral and Brain Sciences*, **19** (1), pp. 107–22.

Chalmers, D. (1995), 'Facing up to the problem of consciousness', *Journal of Consciousness Studies*, **2** (3), pp. 200–19.

Clark, A. (1998), 'Phenomenal consciousness so-called', *Series of Biophysics and Biocybernetics*, Volume 9 (World Scientific: Singapore).

Crick, F. and Koch, C. (1990), 'Towards a neurobiological theory of consciousness', *Seminars in the Neurosciences*, **2**, pp. 263–75.

Cytowic, R. (1993), *The Man Who Tasted Shapes* (New York: Warner).

Damasio, A. (2000), *The Feeling of What Happens* (New York: Harcourt Brace).

Edelman, G. (1989), *The Remembered Present* (New York: Basic Books).

Ellis, R. and Newton, N. (1998), 'Three paradoxes of phenomenal consciousness: Bridging the explanatory gap', *Journal of Consciousness Studies*, **5** (4), pp. 419–42.
Engel, AK, Fries, P., Konig, P., Brecht, M. and Singer, W. (1999), 'Temporal binding, binocular rivalry, and consciousness', *Consciousness and Cognition*, **8** (2), pp. 128–51.
Fauconnier, G. (1985), *Mental Spaces: Aspects of Meaning Construction in Natural Language* (Cambridge, MA: MIT Press).
Gallese, V., Fadiga, L., Fogassi & Rizzolatti, G. (1996), 'Action recognition in the premotor cortex', *Brain*, **119**, pp. 593–609.
Gibson, J.J. (1986), *The Ecological Approach to Visual Perception* (Hillsdale, NJ: Lawrence Erlbaum).
Hardin, L. (2000), 'Red and yellow, green and blue, warm and cool: Explaining colour experience', *Journal of Consciousness Studies*, **7** (8–9), pp. 113–22.
Humphrey, N. (1992), *A History of the Mind* (New York: Springer-Verlag Inc.).
Mack, V. and Rock, I. (1998), *Inattentional Blindness* (Cambridge, MA: MIT Press).
Newton, N. (1989), 'On viewing pain as a secondary quality', *Nous*, **23** (5), pp. 569–98.
Newton, N. (1991), 'Consciousness, qualia and reentrant signalling', *Behavior and Philosophy*, **19**, pp. 21–41.
Newton, N. (1996), *Foundations of Understanding* (Amsterdam: John Benjamins).
Panksepp, J. (1998), *Affective Neuroscience* (New York: Oxford University Press).
Russell, B. (1908), 'Mathematical logic as based on the theory of types', *American Journal of Mathematics*, **30**, pp. 222–62.
Tononi, G. and Edelman, G.M. (1998), 'Consciousness and complexity', *Science*, **282** (5395), pp. 1846–51.

Michael Silberstein

Converging on Emergence
Consciousness, Causation and Explanation[1]

I will argue that emergence is an empirically plausible and unique philosophical/ scientific framework for bridging the ontological gap and the explanatory gap with respect to phenomenal consciousness. On my view the ontological gap is the gap between fundamental ingredients/parts of reality that are not conscious (such as particles and fields) and beings/wholes (such as ourselves) that are conscious. The explanatory gap is the current lack of a philosophical/scientific theory that explains how non-conscious parts can become conscious wholes. Both gaps are of course conceptual as well as empirical in nature. Section I will be devoted to these issues as well as providing other general criteria for an account of consciousness. In section II, different types of emergence will be defined in the context of a more general taxonomy of reduction and emergence. Emergentism about consciousness becomes much more plausible when we see that the ancient 'atomism' (i.e., mereological and nomological supervenience) that drives physicalism on one end, and fundamental property dualism on the other, is probably false. Backing up this claim will be the primary burden of section III. In section IV I will conjecture that phenomenal consciousness is mereologically and perhaps nomologically emergent from neurochemical/ quantum processes, just as many other properties are so emergent. In section V I defend my view of emergence against the objections that: (1) it cannot bridge the explanatory/ontological gap between matter and consciousness and (2) it cannot account for the causal efficacy of consciousness in itself. Finally, in section VI, there is speculation about where all of this might take us in the future.

---

[1] I would like to thank three anonymous referees at *JCS* and Anthony Freeman for their helpful comments. I would like to give a special thanks to Robert Bishop and Harald Atmanspacher for their copious and insightful comments and suggestions.

I: Introduction —
Criteria for a Philosophical Account of Phenomenal Consciousness

1. Defining consciousness

This section provides a yardstick for evaluating my emergentist account in particular as well as competing accounts in general. I wish to provide a philosophical account of phenomenal consciousness (hereafter 'consciousness') that will help lay the foundation for a scientific account of consciousness. Indeed, ideally I think a philosophical account of consciousness should provide a framework for solving the empirical problem of how and why brains are conscious. Specifically, by consciousness I mean: *the experience of being a subject (subjectivity) who experiences qualitative states such as seeing red or feeling pain (so-called qualia)*. Furthermore, it must be explained why/how conscious experience is *unified* or *bound* across short temporal spans such that *we* have simultaneous awareness of diverse features of any conscious state — diverse features that are apparently processed in diverse areas of the brain. I take it as an important open question as to whether or not the notion of qualitative states makes any sense in the absence of a subject. I don't know if it is possible for an *object* (such as the brain) to merely *instantiate* rather than *experience qua subject* qualitative states, but I am doubtful. At any rate, sensory experience, memory, explicit cognition, dreams, emotions as *we* know them all count as consciousness by my definition. Though to be sure, conscious states are otherwise heterogeneous in phenomenal kind and in functional/neural realization.

2. What a theory of consciousness should do

I want my philosophical account of consciousness to answer the following questions:

(1) How is it conceivable/possible that consciousness arises from fundamental elements (such as particles and fields) that are not themselves conscious?
Not just physicalism, but any account of consciousness (including panpsychism/panexperientialism for example) that does not posit *fully conscious experiencing subjects* as fundamental elements of reality (let us call this view *full-blooded idealism*) is going to have to face this question. In my view, to focus on the ontological/explanatory gap between 'matter and mind' creates more confusion than clarity because of the definitional problems involved with both concepts. For example, if we define matter as being inherently non-conscious and we define mind as being essentially conscious then we quickly find ourselves in an inescapable morass, where we are forced into accepting extreme positions or crazy consequences. However for those of us who assume that full-blooded idealism is false (if only for the sake of discussion: after all if full-blooded idealism is true then there are no gaps whatsoever), there is an inescapable ontological/explanatory gap between the fundamental elements of reality (whatever they may be) that happen to be non-conscious and *fully conscious experiencing subjects* such as ourselves. This gap can only be fully bridged by a successful scientific account of how that which is not conscious becomes fully conscious.

The latter ontological/explanatory gap tables the age-old philosophical question about the ultimate *ontic* status of the world in general or of consciousness in particular: is the world or consciousness *essentially physical, mental, functional or neutral in nature*? Though I cannot defend my view fully in this article, I contend that the age-old philosophical question about the ultimate nature of mind, while perhaps meaningful, is irresolvable by either *a priori* or empirical means. More importantly, trying to answer this age-old philosophical question does nothing to help bridge the ontological/explanatory gap between the fundamental elements of reality that happen to be non-conscious and *fully conscious experiencing subjects* such as ourselves. Philosophical accounts of consciousness such as materialism, physicalism, functionalism, panpsychism, etc., *might* provide answers to the age-old philosophical question about the *nature* of consciousness, but they do very little to help bridge the explanatory gap in question, or so I will argue as we go along.

However there is no question that while bridging the explanatory gap in question requires a scientific explanation, that scientific project, like all of science, is inextricably bound with philosophy. On my view science is continuous with philosophy in a number of ways including the obvious historical/causal dimension. First, the scientific method itself and the very practice of science makes the metaphysical assumption that the world is orderly or lawful so as to be induction friendly. The practice of science also presupposes that its methods and our cognitive abilities are up to the task of revealing the deeper truths of the world. Second, any particular scientific theory just is an ontological/epistemological model of the domain it purports to explain. Science just is philosophy of a sort(s). Third, on a more prescriptive note, even philosophy as practised apart from science ought to be informed by our scientific knowledge base at any given time.

One philosophical assumption that I think any scientific account of consciousness must make (or science in general for that matter) is that of monism. Positing monism is a necessary (though not sufficient) condition for bridging the ontological/explanatory gap as well as providing a satisfactory account of 'mental causation'. Though contrary to popular belief I don't think it matters much from the scientific perspective what kind of monism you adopt. Any kind of monism (physicalism, neutral monism, panexperiential monism, etc.) short of full-blooded idealism will leave scientific questions largely untouched and scientific practice unchanged. As Sklar puts it:

> Even if we no longer believed in matter, but perhaps, only in 'ideas in the mind' or 'monads' or 'well-founded phenomena', we could and would still hold on to many of the 'smaller' theories we had constructed earlier on the basis of the presupposition that matter existed. Whether the earth's crust is a portion of a really existing material world or, instead, nothing but a systematically correlated collection of ideas in minds, the issue of whether or not plate tectonics gives a correct account of the dynamical changes in the earth's crustal features remains unaffected (1999, p. 97).

Another well known example of Sklar's point is the fact that Berkeley himself was an atomist. Assuming the falsity of full-blooded idealism, Sklar's point applies equally to the questions of consciousness.

(2) **How is it conceivable/possible that conscious states *qua* consciousness can causally interact with neurochemical states *qua* neurochemical?** Or if you prefer, how is mental causation possible? You might ask, 'Why do you assume that consciousness is causally *efficacious*?' After all, epiphenomenalism itself (non-interactive dualism), and those views such as nonreductive physicalism that imply 'epiphenomenalism', deny the causal *efficacy* of consciousness. As Flanagan notes 'the epiphenomenalist suspicion is extraordinarily hard to dispel' (1992, p. 133). Many people have argued either that whatever causal role consciousness is alleged to play could be accounted for by some purely functional or neural mechanism (Block, 1995) or, that our best physical/scientific accounts of the world actually *preclude* the efficacy of consciousness (Kim, 1993; Levine, 2001). On the other hand, 'It seems overwhelmingly obvious that mental phenomena are both causes and effects of non-mental, physical phenomena' (Levine, 2001, p. 5). Physical phenomena such as light waves cause me to have conscious experiences such as colour experiences, strong emotions cause me to engage in various behaviours, e.g., bodily movements. From the point of view of everyday experience, what could be more certain than the efficacy of consciousness?

I believe in the efficacy of consciousness for three reasons. First, I believe in the efficacy of consciousness on the basis of my first-person conscious experience. Of course first-person conscious experience is not infallible and it's possible that I am simply operating under the *illusion* of causal consciousness. However the notion of epiphenomenalism with respect to consciousness is almost as absurd, disturbing and counterintuitive as outright eliminativism with respect to consciousness. Regarding eliminativism, many of us feel that Descartes wins this one: I feel therefore I feel. From the fact that there seems to be consciousness *it does follow* that there is consciousness, because there is no *seeming* without consciousness! We might be radically deceived about the *nature* of consciousness but not the *fact* of consciousness. The question of eliminativism comes down to this: 'whether first-person conscious experience constitutes pre-theoretic data for a theory of the mind, or is a highly theoretical, and thus epistemically vulnerable posit' (Levine, 2001, p. 137). The vast majority of us who reject eliminativism accept the former. I would argue that our *first-person conscious experience of the efficacy of consciousness* itself constitutes data for a theory of the mind. And as such the belief in the efficacy of consciousness with its high degree of 'certainty' should only be abandoned in the face of more certain countervailing scientific evidence. Many people believe that science has produced such evidence (Levine, 2001; Kim, 1999), but I will argue to the contrary, the best scientific data to date can equally be made to support the efficacy of consciousness. To be fair, there is certainly a disanalogy between eliminativism and epiphenomenalism in that while the former seems impossible in principle, the latter seems merely highly improbable given our first-person conscious experience. There is no *a priori* refutation of epiphenomenalism.

My second reason for believing in the efficacy of consciousness is that it makes naturalizing consciousness much more plausible. The idea of consciousness existing but with no causal capacities is in some ways more disturbing than eliminativism. At least eliminativism (if one can swallow it) has a certain mystery deflating ontological economy, while epiphenomenalism leaves us with the puzzle of how consciousness came to be so ubiquitous in a world in which it makes no difference. Epiphenomenalism makes it very hard to naturalize consciousness by treating it like any other complex biological entity or property from the perspective of the neo-Darwinian paradigm. Consciousness must have causal efficacy in order for it to be treated as an evolutionary adaptation. If consciousness has no effect on fitness then its continued existence cannot be explained by natural selection. While it is true that neo-Darwinian evolution cannot explain the *genesis* of consciousness (Horst, 1999), if consciousness were in fact an adaptation then evolution could at least explain why consciousness continues to flourish. It is possible that consciousness is merely a spandrel, but it is highly improbable given the structural and biological complexity of consciousness, and the amount of bodily energy that goes into creating and maintaining such structures (Grantham and Nichols, 2000). From a neo-Darwinian perspective, the position that consciousness is just a spandrel should only be adopted after all other avenues have been exhausted. As Kim (1993) points out, this is especially true if one holds 'Alexander's Dictum' (to be real is to be causally efficacious) or the 'causal individuation principle' (scientific/natural kinds are truly causal kinds). Given these two principles, to hold that consciousness has no efficacy would be a kind of eliminativism.

My third reason for believing in the efficacy of consciousness comes from the annals of abnormal psychology. While there are many such examples I will focus on 'Conversion Disorder'. Well known conversion symptoms include: 'hysterical' blindness, paralysis, 'anaesthesia', dysphagia, gait disturbance, 'seizure', localized weakness, aphonia, difficulty swallowing, urinary retention, deafness, double vision and (my personal favourite) hallucinations. According to the DSM IV (Diagnostic and Statistical Manual IV) the 'diagnostic features' of conversion disorders are as follows: (1) The presence of symptoms or deficits affecting voluntary motor or sensory function that suggest a neurological or other general medical condition; (2) Psychological factors are judged to be associated with the symptom or deficit, a judgement based on the observation that the initiation or exacerbation of the symptoms or deficit is preceded by conflicts or other stressors. The onset of symptoms occurs very suddenly after the stressful experience that initiates it. The clinical inference often made in such cases is that the patient confronts an acute stressor that creates a psychic conflict, and the physical symptoms serve as the resolution for the conflict; (3) The symptoms are involuntary in that they are not intentionally produced or feigned; (4) The symptoms cannot be fully explained by either a neurological/medical condition or by external causes such as substance abuse or environmental/cultural forces; diagnostic testing shows no physical cause for the dysfunction. Conversion symptoms typically do not conform to known anatomical pathways and physiological mechanisms, but rather follow the patient's conceptualization of a condition. The more medically naïve the person, the more implausible are the presenting symptoms. More sophisticated persons tend to have more subtle symptoms and deficits that may closely simulate neurological or other general medical conditions. A 'paralysis' may involve inability to perform a particular movement or to move an entire body part, rather than a deficit corresponding to patterns of motor innervation. Conversion symptoms are often inconsistent. A 'paralysed' extremity will be moved inadvertently while dressing or when attention is directed elsewhere; (5) The problem must be clinically significant or warrant medical evaluation as evidenced by marked distress; impairment in social, occupational, or other important areas of functioning; (6) The symptoms go beyond pain, sexual dysfunction and are not explicable in terms of other mental disorders such as Somatization Disorder.

Of course if medical science at the ideal limit of its abilities could still detect nothing physically wrong with Conversion Disorder patients then that would be quite a boon for dualism. However I assume, as do most other monists, that as neuroscience grows in technical and diagnostic sophistication it will find underlying neurochemical causal mechanisms or correlates (pick your poison) for Conversion symptoms, just as it did in cases like Clinical Depression. But contra crude physicalism, the fact that Conversion symptoms come with neurochemical mechanisms does in no way negate the primarily *psychological* aetiology of this disorder. Take the following real life example. A Cambodian woman witnesses the torture and murder of her children and other family members. Shortly thereafter the woman, based on her self-report and behaviour, goes blind. Many such Cambodian women with similar symptoms ended up in clinical care in San Francisco in the 1980s and 1990s; they were ultimately diagnosed with Conversion Disorder. It seems clear in such cases that it was the antecedent *psychological* trauma that caused the Conversion symptoms. If the traumatic experience had not occurred, the women would not have gone blind. Furthermore, the women acquired the Conversion symptoms *because* of their conscious (as in consciousness, not in opposition to subconscious) and intentional aspects, such as their love for their children, their sense of guilt at being unable to protect their children and their sense of vulnerability or violation that such horrors are real. To put it crudely, a Conversion Disorder could not befall a zombie.

It is worth pointing out that to date the best treatment for Conversion Disorder is clinical or psychiatric treatment. As I understand it, clinical treatment is reasonably efficacious in this case. Again, this is not an argument for dualism. I don't doubt that Conversion Disorders could also be treated with psychopharmacology. But I do think in the case of Conversion Disorder that a psychopharmacological treatment would only be mitigating the physical causal mechanism (if you will) of the symptoms. In the case of Conversion Disorder the deeper causes derive from our *consciousness per se*.

To summarize, I think an account of consciousness ought to: (1) be realistic about consciousness (i.e., not eliminativist), (2) show how it is conceivable/possible that consciousness arises from fundamental elements that are not themselves conscious and (3) show how is it conceivable/possible that conscious states *qua* consciousness can causally interact with neurochemical states *qua* neurochemical. I will argue throughout that (at least to date) physicalism/ functionalism, panpsychism, etc., fall down on one or more of the three criteria.

II: A Taxonomy of Reduction and Emergence

Because of the equivocation on the terms 'reduction' and 'emergence' in the literature, I will provide a brief taxonomy of reduction and emergence relations that borrows heavily from Van Gulick's (2001 [this volume]) taxonomy, but has been modified in important ways for my own purposes (see Silberstein & Machamer, 2002, for more details). Historically there are two main construals of the problem of reduction and emergence: ontological and epistemological (see Stephan, 1992; McLaughlin, 1992; Kim, 1999, for historical background). The ontological construal of the question: Is there some robust sense in which everything in the world can be said to be *nothing but* the fundamental constituents of reality (such as super-strings) or at the very least, *determined by* those constituents? The epistemological construal of the question: Is there some robust sense in which our scientific theories/schemas (and our common-sense experiential conceptions) about the macroscopic features of the world can be *reduced to* or *identified with* our scientific theories about the most fundamental features of the world? Yet, these two construals are inextricably related. For example, it seems impossible to justify ontological claims (such as the cross-theoretic identity of conscious mental processes with neurochemical processes) without appealing to epistemological claims (such as the attempted intertheoretic reduction of folk psychology to neuroscience) and vice-versa. We would like to believe that the unity of the world will be described in our scientific theories and, in turn, the success of those theories will provide evidence for the ultimate unity and simplicity of the world; things are rarely so straightforward.

Historically 'reductionism' is the 'ism' that stands for the widely held belief that both ontological and epistemological reductionism are more or less true. Epistemological reductionism is the view that the best understanding of a complex system should be sought at the level of the structure, behaviour and laws of its component parts plus their relations. However, according to mereological reductionism, *the relations* between basic parts are themselves reducible to the intrinsic properties of the relata (see below). The ontological assumption implicit is that the most fundamental physical level, whatever that turns out to be, is ultimately the 'real' ontology of the world, and anything else that is to keep the status of real must somehow be able to be 'mapped onto' or 'built out of' those elements of the fundamental ontology. Relatedly, fundamental theory, *in principle,* is deeper and more inclusive in its truths, has greater predictive and explanatory power, and so provides a deeper understanding of the world.

'Emergentism', historically opposed to reductionism, is the 'ism' according to which both ontological and epistemological emergentism are more or less true, where ontological and epistemological emergence are just the negation of their reductive counterparts. Emergentism claims that a whole is 'something more than the sum of its parts', or has properties that cannot be understood in terms of the properties of the parts. Thus, emergentism rejects the idea that there is any fundamental level of ontology. It holds that the best understanding of complex systems must be sought at the level of the structure, behaviour and laws of the whole system and that science may require a plurality of theories (different theories for different domains) to acquire the greatest predictive/explanatory power and the deepest understanding.

It is always possible to divide claims about reductionism and emergentism. One may accept ontological reductionism but reject epistemological reductionism, and vice-versa, likewise for ontological emergentism and epistemological emergentism. Further, one may restrict the question of reductionism and emergentism to particular domains of discourse. For example, one might accept reductionism (or at least one of its construals) for the case of classical mechanics and quantum mechanics, but reject it (or at least one of its construals) for the case of folk psychology and neuroscience.

1. Reduction relations: metaphysical and epistemological

How must things be related for one to ontologically reduce to the other? At least four major answers have been championed:

- Elimination (see Van Gulick, 2001 [this volume], p. 4)
- Identity (see Van Gulick, 2001 [this volume], p. 5)
- Mereological supervenience (includes 'composition', 'realization' and other related weaker versions of this kind of determination relation)
- Nomological supervenience/determination

Mereological supervenience — Reductionism pertaining to parts and wholes goes by several names: 'mereological supervenience', 'Humean supervenience' and 'part/whole reductionism'. Mereological supervenience says that the properties of a whole are determined by the properties of its parts (Lewis, 1986, p. 320).

More specifically, mereological supervenience holds that all the properties of the whole are determined by the qualitative intrinsic properties of the most fundamental parts. Intrinsic properties being non-relational properties had by the parts which these bear in and of themselves, without regard to relationships with any other objects or relationships with the whole. Sometimes philosophers say that intrinsic properties are properties that an object would have even in a possible world in which it alone exists. Paradigmatic examples include mass, charge and spin. Further, intrinsic properties are much like the older primary qualities. It is notoriously difficult to define the notion of an intrinsic property or a relational property in a non-circular and non-question begging manner; nonetheless, philosophers and physicists rely heavily on this distinction (Lewis 1986).

Nomological supervenience/determination — Fundamental physical laws (*ontologically construed*), governing the most basic level of reality, determine or necessitate all the higher-level laws in the universe. Mereological supervenience on the one hand says that the intrinsic properties of the most basic parts *determine* all the properties of the whole — this is a claim about part/whole determination. Nomological supervenience is about *necessity*, the most fundamental laws of physics ultimately necessitate all the special science laws, and therefore these fundamental laws determine everything that happens (in conjunction with initial or boundary conditions). Thus, if two worlds are wholly alike in terms of their most fundamental laws and in terms of initial/boundary conditions, then we should expect them to be the same in all other respects.

Most of the specific variants of epistemological reduction fall into one of four general categories (the last two of which are somewhat modified from Van Gulick's schema):

- Replacement (see Van Gulick, 2001 [this volume], p. 10)
- Theoretical-Derivational (see Van Gulick, 2001 [this volume], p. 10)
- Semantic/model-theoretic/structuralist analysis
- Pragmatic

As Van Gulick notes, the theoretical-derivational account of intertheoretic reduction (Nagel, 1961) was the accepted view among philosophers until very recently. One problem facing the theoretical-derivational account of intertheoretic reduction was forcefully presented by Feyerabend in 'Explanation, Reduction, and Empiricism' (1962). Consider the term 'temperature' as it functions in classical thermodynamics. This term is defined in terms of Carnot cycles and is related to the strict, nonstatistical zeroth law as it appears in that theory. The so-called reduction of classical thermodynamics to statistical mechanics, however, fails to identify or associate *nonstatistical* features in the reducing theory, statistical mechanics, with the nonstatistical concept of temperature as it appears in the reduced theory. How can one have a genuine reduction, if terms with their meanings fixed by the role they play in the reduced theory get identified with terms having entirely different meanings? Classical thermodynamics is not a statistical theory. The very possibility of finding a reduction function or bridge law that captures the concept of temperature and the strict, nonstatistical role it plays in the thermodynamics seems impossible.

Many physicists, now, would accept the idea that our concept of temperature and our conception of other exact terms that appear in classical thermodynamics, such as 'entropy', need to be reformulated in light of the alleged reduction to statistical mechanics. Textbooks, in fact, typically speak of the theory of 'statistical thermodynamics'.

Because of the problem mentioned above, as well as others, many philosophers of science felt that the theoretical-derivational model (Nagel, 1961) did not realistically capture the actual process of intertheoretic reduction. As Primas puts it, 'there exists not a single physically well-founded and nontrivial example for theory reduction in the sense of Nagel (1961). The link between fundamental and higher-level theories is far more complex than presumed by most philosophers' (1998, p. 83). Therefore, alternative models of intertheoretic reduction abandon one or more of the following assumptions made by the theoretical-derivational account (i.e., the logical empiricist account):

Ontological assumptions

(1) Property/kind cross-theoretic (ontological) identities are to be determined solely by formal criteria such as successful intertheoretic reduction, e.g.,

smooth intertheoretic reduction is both necessary and sufficient for cross-theoretic identity.
(2) Realism, scientific theories are more than mere 'computational devices'.

Epistemological assumptions

(1) Philosophy of science is prescriptive rather than descriptive, e.g., philosophy of science should seek a grand, universal account of intertheoretic reduction.
(2) Scientific theories are axiomatic systems.
(3) Reduction = logical deduction, or at least deduction of a structure specified within the vocabulary and framework of the reduced theory or some corrected version of it.
(4) Necessity of bridge laws or some other equally strong cross-theoretic connecting principles to establish synthetic identities.
(5) Symbolic logic is the appropriate formalism for constructing scientific theories.
(6) Scientific theories are linguistic entities.
(7) Hardcore explanatory unification. Reduction is proof of displacement (in principle) showing that the more comprehensive reducing theory contains explanatory and predictive resources equaling or exceeding those of the reduced theory.
(8) Intertheoretic reductions are an all or nothing *synchronic* affair as in the case of 'microreductions' (Oppenheim and Putnam, 1958; Causey, 1977): the lower-level theory and its ontology reduce the higher-level theory and its ontology. Ontological levels are mapped one to one onto levels of theory which are mapped one to one onto fields of science.
(9) The architecture of science is a layered edifice of analytical levels (Wimsatt, 1976).

Alternatives to the Nagel model (1961) are deemed more or less radical (by comparison) depending on which of the preceding tenets are abandoned. On the more conservative side, many alternative accounts of intertheoretic reduction merely modify (3) by moving to logico-mathematical deduction, but reject (4). For example, the requirement of bridge laws gets replaced by notions such as: 'analog relation' — an ordered pair of terms from each theory (Hooker, 1981; Bickle, 1998), 'complex mimicry' (Paul Churchland, 1989) or 'equipotent image' (Patricia Churchland, 1986), to name a few. Many of these comparatively conservative accounts also reject (8), preferring to talk about a range of reductions, from replacement on one end of the continuum to identity on the other. More radical alternatives to the Nagel model are as follows.

Semantic/model-theoretic/structuralist analysis: This approach (the 'semantic' approach for short), is regarded by some as comparatively radical because it rejects the conception of scientific theories as formal calculi formalizable in first-order logic and (partially) interpretable by connecting principles such as bridge laws. The semantic approach makes the following assumptions: (i) Scientific theories are not essentially linguistic entities (sets of sentences), but are terms or families of their *mathematical models or mathematical structures* and (ii) The formal explication of the structure of scientific theories is not properly carried out with first-order logic and metamathematics, but with *mathematics*, though the choice of mathematical formalisms will differ depending on who you read (Giere, 1988; Bickle, 1998; Batterman, 2000). The semantic approach minimally rejects epistemological assumptions (2)–(6) and (8), i.e., rejects the logical derivation of laws and abandons truth preservation (everything the reduced theory asserts is also asserted by the reducing theory). On the semantic approach the reduction relation might be conceived of as some kind of 'isomorphism' or 'expressive equivalence' between models (Bickle, 1998). However, as we shall shortly see, more radical versions of the semantic approach reject all the preceding epistemological assumptions held by the logical empiricist account of intertheoretic reduction.

Pragmatic: Success in real world representation is in large part a practical matter of whether and how fully one's attempted representation provides *practical causal* and *epistemic access* to the intended representational target. A good theory or model succeeds as a representation if it affords reliable avenues for *predicting*, *manipulating* and *causally* interacting with the items it aims to represent. It is the practical access that the model affords in its context of application that justifies viewing it as having the representational content that it does (Van Fraassen, 1989; Kitcher, 1989). If a lower-level theory about a specific domain provides superior *real-world explanatory* and *predictive* value compared to a higher-level theory representing the same domain, then the lower-level theory has met the ultimate test of successful intertheoretic reduction. Note that this contextual, pragmatic account of intertheoretic reduction is also highly *particularist;* it advocates adjudicating on a case-by-case basis; no universal theory of reduction is sought. This account rejects at least assumptions (1)–(6) in the epistemological category, and assumption (1) in the ontological category (Patricia Churchland, 1986). Though again, more radical versions reject all nine of the preceding epistemological assumptions.

Whereas the theoretical-derivational account (i.e., the logical empiricist account) of intertheoretic reduction (and its variants) only makes sense if you presuppose nomological and mereological supervenience, in principle, both the semantic and the pragmatic accounts of intertheoretic reduction are compatible with mereological emergence and perhaps even nomological emergence (defined below). We shall encounter specific versions of such accounts of reduction shortly.

2. Emergence relations: metaphysical and epistemological

At least four major forms of ontological emergence have been championed; each is an elaboration of the failure of its corresponding reduction relation:

- Non-elimination
- Non-identity
- Mereological emergence (holism)
- Nomological emergence

Non-elimination: If a property, entity, causal capacity, kind or type cannot be eliminated from our ontology, then one must be a realist about said item. Obviously, this leaves open the question of what the criteria ought to be for non-elimination in any given case; but they will almost certainly be epistemological/explanatory in nature.

Non-identity: If a property, type or kind cannot be ultimately identified with a physical (or lower-level) property, type or kind then one must accept that said item is a distinct non-physical (or higher-level) property, type or kind. Again, this leaves open the criteria for non-identifiability and again, such criteria are generally epistemological/explanatory in nature.

Mereological emergence (holism): These are cases in which objects have properties that are not determined by the intrinsic (non-relational) physical properties of their most basic physical parts. Or, cases in which objects are not even wholly composed of basic (physical) parts at all. British (classical) emergentism held that mereological emergence is true of chemical, biological and mental phenomena (McLaughlin, 1992).

Nomological emergence: These are cases in which higher-level entities, properties, etc., are governed by higher-level laws that are not determined by or necessitated by the fundamental laws of physics governing the structure and behaviour of their most basic physical parts. For example, according to Kim (1993), British emergentism held that while there were bridge laws linking the biological/mental with the physical, such bridge laws were inexplicable brute facts. That is, on Kim's view British emergentism did not deny global supervenience. But British emergentism did deny that the laws governing the mental for example were determined by (or explained by) the fundamental laws of physics (Kim, 1993; McLaughlin, 1992). In other words, British emergentism held that it is simply a brute fact that consciousness emerges given particular basal conditions. A more severe example of nomological emergence would be where there were no bridge laws whatsoever linking fundamental physical phenomena with higher-level phenomena. In such cases fundamental physical facts and laws would only provide a necessary condition for higher-level facts and laws. This would imply possible violations of global supervenience. Both Cartwright (1999) and Dupré (1993) *seem* to defend something like this kind of nomological emergence. An even more severe example is found in cases in which either fundamental physical phenomena or higher-level phenomena are not law-governed at all. This would amount to eliminativism or antirealism regarding nomological or physical necessity (see Van Frassen, 1989, for a defence of this view). It is important to note that in all cases of *nomological emergence*, it is *in principle* impossible to derive or predict the higher-level phenomena on the basis of the lower-level phenomena.

As Van Gulick says, and it's worth repeating in this case, at least two major views have been championed regarding epistemological emergence:

- Predictive/Explanatory Emergence
- Representational/Cognitive Emergence

Predictive/Explanatory Emergence: Wholes (systems) have features that cannot *in practice* be explained or predicted from the features of their parts, their mode of combination, and the laws governing their behaviour. In short, X bears predictive/explanatory emergence with respect to Y if Y cannot (reductively) explain X. More specifically, in terms of types of intertheoretic reduction, X bears predictive/explanatory emergence with respect to Y if Y cannot *replace* X, if X cannot be *derived* from Y, or if Y cannot be shown to be *isomorphic* to X. A lower-level theory Y (description, regularity, model, schema, etc.), for purely epistemological reasons (conceptual, cognitive or computational limits), can fail to predict or explain a higher-level theory X. If X is predictive/explanatory emergent with respect to Y for *all possible cognizers in practice*, then we might say that X is *incommensurable* with respect to Y. A paradigmatic and notorious example of predictive/explanatory emergence is chaotic, non-linear dynamical systems (Silberstein and McGeever, 1999). The emergence in chaotic systems (or models of non-linear systems exhibiting chaos) follows from their sensitivity to initial conditions, plus the fact that physical properties can only be specified to finite precision; infinite precision would be necessary to perform the required 'reduction', given said sensitivity. It does not follow, however, that chaotic systems provide evidence of violations of mereological supervenience or nomological supervenience (Kellert, 1993, p. 62 and p. 90). e.g., dynamical systems have attractors as high-level emergent features only in the sense that you cannot deduce them from equations for the system. McGinn (1999) and other mysterians hold that folk psychology is predictive/explanatory emergent with respect to neuroscience.

Representational/Cognitive Emergence: Wholes (systems) exhibit features, patterns or regularities that cannot be fully represented (understood) using the theoretical and representational resources adequate for describing and understanding the features and regularities of their parts and (reducible) relations. X bears representational/cognitive emergence with respect to Y, if X does *not* bear predictive/explanatory emergence with respect to Y, but nonetheless X represents higher-level patterns or non-analytically guaranteed regularities that cannot be fully, properly or easily represented or understood from the perspective of the lower-level Y. As long as X retains a significant *pragmatic* advantage over Y with respect to understanding the phenomena in question, then X is representational/cognitive emergent with respect to Y. Nonreductive physicalism holds that folk psychology is representational/cognitive emergent with respect to neuroscience (Antony, 1999).

III: Emergence and Physics

1. Quantum holism (superposition and entanglement) and mereological emergence

Before I say more about consciousness as an emergent property, by way of analogy, I will give examples of other properties that many believe are emergent in just the sense defined above. There are some people who allege that *quantum mechanics itself* provides examples of mereological emergence:

> In quantum theory, then, the physical state of a complex whole cannot always be reduced to those of its parts, or to those of its parts together with their spatiotemporal relations, even when the parts inhabit distinct regions of space. Modern science, and modern physics in particular, can hardly be accused of holding reductionism as a central premise, given that the result of the most intensive scientific investigations in history is a theory that contains an ineliminable holism (Maudlin, 1998, p. 55).

By and large, a system in classical physics can be analysed into parts, whose states and properties determine those of the whole they compose. But the state of a system in quantum mechanics resists such analysis. The quantum state of a system gives a specification of its probabilistic dispositions to display various properties on its measurement. Quantum mechanics' most complete such specification is given by what is called a pure state. Even when a compound system has a pure state, its subsystems generally do not have their own pure states. Schrödinger, emphasizing this characteristic of quantum mechanics, described such component subsystems as 'entangled'. Such entanglement of systems demonstrates nonseparability — the state of the whole is not constituted by the states of its parts. State assignments in quantum mechanics have been taken to violate state separability in two ways: the subsystems may simply not be assigned any pure states of their own, or else the states they are assigned may fail to completely determine the state of the system they compose.

On the basis of nonseparability, many people have argued that quantum mechanics provides us with examples of systems with properties that do not always reduce to the intrinsic properties of the most basic parts, i.e., quantum mechanical systems exhibit mereological emergence (Hawthorne and Silberstein, 1995; Humphreys, 1997; Healey, 1991). Such entangled systems appear to have novel properties of their own. Quantum systems that are in superpositions of possible states are behaviourally distinct from systems that are in mixtures of these states and individual systems can be become entangled and thus form a new unified system which is not the sum of its intrinsic parts. From this some further infer that: 'the state of the compound [quantum] system determines the state of the constituents, but not vice versa. This last fact is exactly the reverse of what [mereological] supervenience requires' (Humphreys, 1997, p. 16).

The opinion of a growing number of philosophers of physics is expressed by Maudlin as follows:

> Quantum holism ought to give some metaphysicians pause. As has already been noted, one popular 'Humean' thesis holds that all global matters of fact supervene on local matters of fact, thus allowing a certain ontological parsimony. Once the local facts have been determined, all one needs to do is distribute them throughout all of space–time to generate a complete physical universe. Quantum holism suggests that our world just doesn't work like that. The whole has physical states that are not determined by, or derivable from, the states of the parts. Indeed, in many cases, the parts fail to have physical states at all. The world is not just a set of separately existing

localized objects, externally related only by space and time. Something deeper, and more mysterious, knits together the fabric of the world. We have only just come to the moment in the development of physics that we can begin to contemplate what that might be (1998, pp. 58–60).

At any rate, quantum nonseparability is not restricted to settings such as twin-slit experiments and EPR (non-locality) experiments. Superpositions and entangled states are required to explain certain chemical and physical phenomena such as phase transitions that give rise to superconductivity, superfluidity, paramagnetism, ferromagnetism (see Anderson, 1994; Auyang, 1998; Cornell and Wieman, 1998).

Some interpretations of quantum mechanics such as Bohr (1934) and Bohm (Bohm & Hiley, 1993) imply mereological emergence (holism) with respect to *entities*: there are physical objects that are not wholly composed of basic (physical) parts. On Bohr's interpretation one can meaningfully ascribe properties such as position or momentum to a quantum system only in the context of some well-defined experimental arrangement suitable for measuring the corresponding property. Although a quantum system is purely physical on this view, it is not composed of distinct happenings involving independently characterizable physical objects such as the quantum system on the one hand, and the classical apparatus on the other. On Bohm's interpretation, it is not just quantum object and apparatus that are holistically connected, but any collection of quantum objects by themselves constitute an indivisible whole. A complete specification of the state of the 'undivided universe' requires not only a listing of all its constituent particles and their positions, but also of a field associated with the wave-function that guides their trajectories. If one assumes that the basic physical parts of the universe are just the particles it contains, then this establishes ontological holism in the context of Bohm's interpretation.

2. Symmetry breaking and mereological emergence

Our current best *fundamental theory of matter* (the so-called standard model) appears to have mereological emergence built right into it. Paradigmatic examples of intrinsic properties include particle mass, charge and spin. However our best fundamental physical theory regarding such properties tells us that they are inherently relational, requiring irreducible interactions for their existence as well as particular values.

Theories of everything (TOEs) seek to combine all physical laws and principles, some even into a single, simple formula. From this would flow a correct description of all the forces of nature, all the basic particles and fields from which the universe is composed. Such a theory would allegedly show that there is only one fundamental force in the universe that has come to display itself as if it were four different forces; or, at least, that at some point in the very distant past the four forces were one force, but became successively distinct from one another as the universe 'cooled off' after the Big Bang. For example, according to GUT (Grand Unified Theories in which gravity is excluded), the energy scale at which the unification of all three particle forces takes place is enormously large — just below the Planck energy. Near the instant of the Big Bang, where such energies were found, the theory predicts that all three particle forces were unified by one GUT symmetry. As the universe rapidly 'cooled down', the original GUT symmetry was successively 'spontaneously' broken into the symmetries of the strong and electroweak forces (Kaku, 1993).

This project of unification faces a dilemma however. As Maudlin puts it:

> Obviously, manifest symmetry of the forces is precluded by the spontaneous symmetry breaking. Unification is to be sought in spite, rather than because, of the immediately observable properties of the forces. The mechanism of symmetry breaking allows the research program to continue in the face of the apparent dissimilarity of the forces, but it also denies us direct empirical grounds for believing that there is any hidden symmetry at all (1996, p.141).

Regarding spontaneous symmetry breaking of the four forces themselves, according to the standard model one 'force' becomes four irreducible forces as the universe 'cools down' (though this is overly simplistic as the notion of 'temperature' implied in this case means much more than 'temperature' in standard thermodynamics, Kaku, 1993). Or more precisely, the four forces emerge as the result of cosmological 'phase transitions'. The spontaneous symmetry breaking mechanism via phase transitions explains how one thing becomes four very different things. The symmetry breaking of the four forces might appear so far to be a process of emergence that requires nothing outside the system to occur.

However it turns out that a very striking kind of mereological emergence *à la* spontaneous symmetry breaking is required to explain the diverse and essential properties of different fundamental ('force carrying') particles. Maudlin's dilemma is this: How can we unify fundamental particles and still maintain that they are intrinsically different? The mechanism of spontaneous symmetry breaking in the standard model is an attempt to resolve this dilemma. So why are the fundamental particles different? According to the standard model, prior to spontaneous symmetry breaking, the particles that will become electrons and neutrinos are identical or symmetric and that symmetry must be broken in order for them to come to possess different properties such as their different values for mass. The symmetry breaking in question is provided by the Higgs field. More specifically, the particles that will become electrons and neutrinos interact with Higgs particles (a particular kind of Higgs particle call it the 'electron-Higgs'), but because of the kind of Higgs particle involved in such interactions, only a relatively small number of the antecedent symmetrical particles will become the comparatively mass-less neutrinos. Because the actual universe contains only the presence of electron-Higgs, the symmetry between electrons and neutrinos is broken in favour of electrons being much more massive than neutrinos. The laws of physics say that the electron and the neutrino are symmetrical, but such symmetries are unstable, thus, they must be broken one way or the other in favour of more stable asymmetric configurations. The laws of physics are symmetric, but the only stable configurations are asymmetric. In this case the symmetry of the laws has been spontaneously broken by the Higgs field (the electron-Higgs field in particular) permeating the whole of space.

The spontaneous symmetry breaking mechanism (in some form or another) is supposed to explain in general why the fundamental particles have the various properties that they do as well as solve the 'hierarchy problem' (the problem of explaining why the four forces have such vastly different and unexplained strengths) — for example why gravity is so weak with respect to the other forces. The standard model posits Higgs particles to distinguish between weak interactions and electromagnetism and to distinguish those from the strong interactions. Whether or not the world is filled by a gas of Higgs particles depends on conditions such as 'temperature'. Higher 'temperatures' mean unstable configurations can be maintained by thermal energy. In some of the universe's phase transitions Higgs particles can exist, their type and ratio presumably being a function of the overall temperature and density of matter. The other well known solutions for resolving the hierarchy problem, the problem of determining fundamental physical values such as mass, etc., involve equally *non-intrinsic mechanisms* such as partner particles posited by supersymmetry (e.g., 'quarks' and 'squarks') or extra dimensions posited by string theory.

3. Key features of quantum holism and symmetry breaking

There are several things worth mentioning about superposition/entangled states and symmetry breaking:

(1) No new extra ingredients or fundamentals, 'configurational forces', vital entelechies, etc., need be added to explain such properties, hence monism is not violated.
(2) Yet, neither are such properties determined by or mereologically supervenient upon more basic properties.
(3) Such properties are relatively abstract (i.e., non-concrete). Such properties are not substances or intrinsic properties, but rather they are irreducible states or relations. In some cases it is not even clear that such properties have proper physical parts or a definite spatiotemporal location (they are probability distributions [a wave function] 'spread-out' over a region of space–time).
(4) Such properties are unquestionably causally efficacious; neither physics nor chemistry and their explanatory schemes makes sense otherwise.
(5) Such properties give us reason to doubt that the ordering on the complexity of structures ranging from those of elementary physics to those of astrophysics and neurophysiology is discrete. But even so, the interactions between such structures will be so entangled that any separation into levels will be conventional or contextual (Humphreys, 1997; Silberstein, 1998; Silberstein & McGeever, 1999). Reality may not divide into a discrete hierarchy of levels. For example, in entangled quantum systems the newly resulting 'compound' determines the prior and original constituents (the particles) and not the other way around as mereological supervenience claims. The point is that the physical and mental may not form a dichotomy at all, but rather a continuum. Thus the mental and the physical may only be conventionally or contextually separable from one another. More generally, the microscopic and the

macroscopic may only be contextually separable from one another. Such properties warn us not to reify the layered conception of the world. The standard divisions and hierarchies between phenomena that are considered fundamental and emergent, simple and aggregate, kinematic and dynamic, and perhaps even what is considered physical, biological and mental are redrawn and redefined. Such divisions will be dependent on what question is being put to nature and what scale of phenomena is being probed. It is true that science is divided into hierarchical descriptions and theories, but given mereological emergence, this might be only an epistemological artifact of scientific explanatory practice and not a fact about the world. The fact that current psychology does not deal with properties like spin and current physics does not mention mental properties like pain does not mean that so-called mental properties could not be quantum mechanical in nature.

(6) Perhaps most importantly of all, the kinds of explanations of such properties and involving such properties *does not require and does not allow* that the phenomena being explained be explicable or describable (solely) in the vocabulary of the underlying theory or in terms of the most basic parts.

4. Physics and possible nomological emergence

Nor does ancient atomism seem to be fairing any better when it comes to nomological supervenience: 'Our scientific understanding of the world is a patchwork of vast scope; it covers the intricate chemistry of life, the sociology of animal communities, the gigantic wheeling galaxies, and the dances of elusive elementary particles. But it is a patchwork nevertheless, and the different areas do not fit well together' (Berry, 2000, p. 3). Regarding nomological emergence, a growing body of literature focusing on actual scientific practice suggests that there really are not many cases of successful intertheoretic reduction in the empiricist tradition of *demonstrating* nomological supervenience . Specific cases for which this claim has been made include:

(1) the reduction of thermodynamics to statistical mechanics (Primas, 1991; 1998; Sklar, 1999);
(2) the reduction of thermodynamics/statistical mechanics to quantum mechanics (Hellman, 1999);
(3) the reduction of chemistry to quantum mechanics (Cartwright, 1997; Primas, 1983);
(4) the reduction of classical mechanics to quantum mechanics (such as the worry that quantum mechanics cannot recover classical chaos, Belot and Earman, 1997).

Take the case of chemistry and its alleged reduction to quantum mechanics. Currently chemists do not use fundamental quantum mechanics (Hamiltonians and Schrödinger's equation) to do their science. Quantum chemistry cannot be deduced directly from Schrödinger's equation due to multiple factors that include the many-body problem (Hendry, 1998). Quantum mechanical wave functions are not well-suited to represent chemical systems or support key inferences essential to chemistry (Woody, 2000). It is still an open question as to whether or not quantum mechanics can describe or represent a molecule (Berry, 2000). Indeed, little of current chemistry can be represented by pure quantum mechanical calculations (Scerri, 1994; Ramsey, 1997; Primas, 1983). Chemistry uses idealized models whose relationship to fundamental quantum mechanics is questionable (Hendry, 1999; Primas, 1983). As Cartwright puts it: 'Notoriously, we have nothing like a real reduction of the relevant bits of physical chemistry to physics — whether quantum or classical. Quantum mechanics is important for explaining aspects of chemical phenomena but always quantum concepts are used alongside of *sui generis* — that is, unreduced — concepts from other fields. They do not explain the phenomena on their own' (1997, p. 163,).

Another well known example is the case of thermodynamics and statistical mechanics. First, there is a variety of distinct concepts of both temperature and entropy that figure in both statistical mechanics and classical thermodynamics. Second, thermodynamics can be applied to a number of very differently constituted microphysical systems. Thermodynamics can be applied to gases, electromagnetic radiation, magnets, chemical reactions, star clusters and black holes. As Sklar puts it, 'The alleged reduction of thermodynamics to statistical mechanics is another one of those cases where the more you explore the details of what actually goes on, the more convinced you become that no simple, general account of reduction can do justice to all the special cases in mind' (1993, p. 334). Third, the status of the probability assumptions that are required in order to recover thermodynamic's principles within statistical mechanics are themselves problematic or *ad hoc*. For example, the assumption that the micro-canonical ensemble is to be assigned the standard, invariant, probability distribution. Fourth, perhaps the thorniest problem of all, statistical mechanics is time symmetric and thermodynamics possesses time asymmetry.

Recent accounts of *intertheoretic reduction*, the more radical versions of the semantic and pragmatic models I mentioned earlier, such as *GRR* (Schaffner, 1992; 1998) and the more explicitly pragmatic and ontic *causal mechanical* model (Machamer *et al.*, 2000), explicitly reject microreduction, in part because of the cases just mentioned. The causal mechanical model of intertheoretic reduction focuses on explanations as characterizing complex (nested and interconnected) causal mechanisms and pathways, such as we find in molecular biology and neuroscience. The emphasis in this model is on causal/mechanical processes as opposed to nomological patterns of explanation. More importantly for our purposes, this model admits of *multilevel* descriptions of causal mechanisms that mix different levels of aggregation from cell to organ back to molecule.

Take the following example from behavioural genetics: 'there is no *simple* [reductive] explanatory model for behavior even in simple organisms. What *C. elegans* [a simple worm] presents us with is a tangled network of influences [causal mechanisms] at genetic, biochemical, intracellular, neuronal, muscle cell, and environmental levels' (Schaffner, 1998, p. 237). This kind of reductive explanation focuses on interlevel causal processes and emphasizes the limits and rarity of logical empiricist accounts of intertheoretic reduction. This approach to reduction is diachronic, emphasizing the gradual, partial and fragmentary nature of many real world cases. This model also views intertheoretic reduction as a continuum and not a dichotomy.

One can also find similar 'web-like' and 'bushy' cases of intertheoretic reduction within physics. For example, cases in which two domains (such as quantum mechanics and chemistry) are related by an asymptotic series often require appeal to an intermediate theory (Batterman, 2000; 2001; Primas, 1998; Berry, 1994). In the asymptotic borderlands between such theories phenomena emerge that are not fully explainable in terms of either the lower-level or the higher-level theory, but require both theories or an intermediary (Batterman, 2000). Examples of this phenomena can be found in the borders between quantum mechanics and chemistry, as well as thermodynamics and statistical mechanics (Berry, 1994; 2000; Berry & Howls, 1993; Batterman, 2000). Batterman speaks of the 'asymptotic emergence of the upper level properties' in such cases, and he goes on to suggest that 'it may be best, in this context, to give up on the various philosophical models of reduction which require the connection of kind predicates in the reduced theory with kind predicates in the reducing theory. Perhaps a more fruitful approach is to investigate asymptotic relations between the different theory pairs. Such asymptotic methods often allow for the understanding of emergent structures which dominate observably repeatable behavior in the limiting domain between the theory pairs' (2000, pp. 136–7). Intertheoretic reduction à la singular asymptotic expansions is not easy to characterize, though perhaps it is fair to say that it has elements of both nomological and causal explanation. Examples of intertheoretic relations involving singular asymptotic expansions include: Maxwell's electrodynamics and geometrical optics; molecular chemistry and quantum mechanics and; classical mechanics and quantum mechanics (Primas, 1998; Berry, 2000).

There are several things worth noticing about both the preceding models of intertheoretic reduction. Such reductions are not universally valid, they can only be considered on a case by case basis. Such reductions require specification of context, the new description or higher-level theory cannot be derived from the lower-level theory. Indeed, such reductions generally start with the higher-level theory/context and work back to the more fundamental theory (Berry, 1994). The lower-level theory (the reducing theory) is not, as a rule, more powerful or universal in its predictive/explanatory value than the higher-level theory (the reduced theory). Indeed, the new ontology and topology generated by the higher-level description cannot be replaced or eliminated precisely because of its more universal explanatory power; and the intertheoretic reductions on such accounts show why this must be the case. Contrary to the standard view, failure of reduction need not imply failure of explanation. A more fundamental theory can explain a higher-level theory ('from below' as it were) without providing a reduction of that theory in the standard senses of the term. Emergent phenomena need not be inexplicable brute facts contrary to classical emergentism. Given such accounts of intertheoretic reduction, there is good reason to think that contra the dreams of the unity of science movement, that unification of scientific theories will be local at best.

IV: Consciousness and Emergence

If mereological and nomological emergence exists *within fundamental physics itself,* then it is possible that consciousness might be an emergent property in just these senses. My conjecture is that consciousness is mereologically and perhaps nomologically emergent with respect to its neurochemical base. By analogy, perhaps all of the preceding points about mereological and nomological emergence within physics are true of consciousness (and its explanation) as well. Assuming that they are goes some distance toward meeting the criteria given for a scientifically respectable philosophical account of consciousness. This assumption might also provide an answer to Van Gulick's lament about the prospects for explaining consciousness: 'surely there must be intermediate cases that involve explanatory rather than merely brute links, but nonetheless fall short of the apriorists' radical requirement for strict logical entailments' (Van Gulick, 2001 [this volume], p. 12). Emergence might be a middle path between such extremes.

Consciousness is a qualitatively new systemic property of the whole that possesses causal properties that are not reducible to any of the intrinsic causal properties of the most basic parts; consciousness is either an irreducible relational property of the relata or it is a superposition/entanglement-like property that denies or subsumes the autonomous existence of the relata. Consciousness either subsumes the intrinsic properties of the basic parts or exerts causal influence on the basic parts, consistent with but distinct from the causal properties of the basic parts themselves (Silberstein, 1998, p. 468).

If consciousness is an emergent property as hypothesized, what kind of explanation should we look for specifically and where will we find it? This empirical question is the primary challenge for the kind of emergentism about consciousness that I am advocating. The closest existing accounts of consciousness to the kind of emergentism I am suggesting fall into two basic categories: non-quantum (classical) emergentists and quantum emergentists (Silberstein, 1998; Silberstein & McGeever, 1999).

Contra the quantum emergentists, the non-quantum emergentists hold one of the following: (1) quantum mechanics is not a theory of everything, e.g., it does not apply to brains/minds, (2) quantum mechanics is a theory of everything, but that quantum mechanics itself explains why quantum effects (such as entangled states) are screened off from the classical world, or (3) quantum mechanics might be a theory of everything, but the quantum mechanics of brains, like much of molecular and chemical quantum mechanics, is mathematically, computationally or cognitively intractable. Quantum emergentists of course deny 1–3 (for details on quantum accounts of consciousness and their problems see Silberstein, 1998).

The non-quantum emergentists seek an explanation for the emergence of consciousness in the classical world. One can find such explanations within theoretical cognitive neuroscience, the body of mathematics known as non-linear dynamics, chaos theory or the field of complexity studies. This approach to the brain/mind is sometimes called the dynamical systems approach:

> The dynamical systems approach for understanding the brain holds that emergent patterns driven by weak local correlations are the proper currency of cognition and perception [the unit of thought]. Common 'carrier' waves emerge from the background noise of a sea of immense activity (Hardcastle, 1999, pp. 81–2).

The dynamical systems approach rejects the classic view that the firing neuron is the basic brain unit. What motivates this approach? Allow me to quote at length:

> The impetus for adopting a dynamical systems perspective in the brain sciences comes from several quarters. First, it is clear that information in the brain is transmitted by far more than action potentials and neurotransmitters. Hormones and neuropeptides impart data through the extracellular fluid more or less continuously in a process known as 'volume transmission.' What is important is that these additional ways of communicating among cells in the central nervous system mean that simple (or even complicated) linear or feedforward models are likely to be inaccurate. The model of the brain as a serial processing computer ignores much important computation and communication in the head. Discovering the importance of global communication in the brain has led some to conclude that it is better to see our brain as a system that works together as a complex interactive whole for which any sort of reduction to lower levels of description means a loss of telling data. Moreover, neurons respond differently when signals arrive simultaneously than when they arrive merely close together. This electrical oscillation affects how the neuron can respond to other inputs. And small changes in the frequency of oscillation (caused by changing the input signals ever so slightly) can produce large changes in a neuron's output pattern (Hardcastle, 1999, pp. 78–82).

Many researchers these day agree that coherent neural activity (in some form or another) is the mechanism by which the brain realizes consciousness (Revonsuo, 2001). In general such a mechanism will involve synchronous neural activity at high frequencies. Some researchers emphasize complex electrophysiological activity in thalamocortical loops and subcortical circuits, realized by synchronous neural firing 'scanning' the cortex with high frequencies; or synchronous oscillatory activity in the cortex (Revonsuo, 2001, p. 6). Others emphasize large-scale, complex electrophysiological or bioelectrical activity patterns involving millions of neurons and billions of synapses. Such large-scale patterns or neural networks be would spatially distributed in the brain. Such patterns (e.g., synchronous oscillations or some other mechanism) would be coherent or unified but would nonetheless change constantly and very rapidly (Revonsuo, 2001, p. 7).

To be sure, both camps of emergentists have to face Herculean efforts to overcome Herculean barriers. Interpretational issues aside, quantum emergentists must explain how and why, contrary to the best evidence at the moment (Tegmark, 2000; Seife, 2000), quantum effects could be active (not screened-off) at the level of neurochemical processes. Non-quantum emergentist explanations of consciousness, on the other hand, generally appeal to the non-linearity, chaos and/or 'complexity' of brain processes (Silberstein & McGeever, 1999). The problem here is that according to the standard wisdom of classical physics in general and non-linear dynamical systems theory/chaos theory in particular, non-linear systems (chaotic or otherwise) are typically treated as being 'deterministic' in two ways. First classical systems, non-linear or linear, have 'deterministic' or definite values at all times (e.g., no superpositions or entangled states) and second such systems evolve in a purely deterministic (non-stochastic) Laplacian fashion (Silberstein & McGeever, 1999; Crutchfield, 1994; Kellert, 1993). It is hard to imagine how any system that is 'deterministic' in both these ways could exhibit mereological emergence. And as classically conceived, brain processes are just such deterministic systems. Of course classical mechanics, non-linear dynamical systems theory, etc., are open to interpretation and reformulation just as quantum mechanics is. There are attempts to reinterpret (or reformulate) these theories to allow for truly stochastic processes. There are also attempts to combine quantum and chaotic effects. For example, there are those who argue that classical level (mesoscopic) quantum effects can be amplified via chaotic sensitive dependence to initial conditions (Kellert, 1993; Bishop & Kronz, 1999).

V: Answering Objections

Standard objections to the view I have been espousing are as follows:

(1) Emergence cannot account for causally efficacious consciousness (mental causation) without violating our best account of physicalism, physics or both.
(2) Emergence cannot bridge the ontological/explanatory gap.

Objection # 1: Emergence and the efficacy of consciousness

Several people have alleged that in order for emergent consciousness qua consciousness to be truly efficacious it would have to engage in 'downward causation', a kind of causation which is allegedly impossible or incoherent from the perspective of physicalism and/or physics (e.g., Kim, 1993; 1998; 1999). The argument is based on three key assumptions. First, the realization principle, which says that each mental event/property must be realized by, determined by or supervene upon some physical event/property. Second, the causal closure of the physical domain: 'the domain of physical phenomena, according to current physical theory, is causally closed. Nothing can affect the distribution of matter and energy in spacetime except the instantiation of basic properties in basic objects that occupy spacetime' (Antony & Levine, 1997, p. 100). Thus, causal closure says that every physical event in the universe is caused by some other physical event. Third, the causal inheritance principle, which says that every higher-level property P inherits its causal capacities from its lower-level realizer R.

Given these assumptions, in order for a mental event M to causally bring about mental event M*, M would have to cause M*'s physical realizer P* to come into being. As Kim puts it, 'there can be no same level causation without cross level [downward] causation' (1996, p. 200). But of course event M is itself realized by event P, and given the causal closure of the physical, P is sufficient to bring about P*. Therefore, positing causal consciousness leads to causal over-determination (which is absurd) and violations of causal closure. Because each conscious state qua consciousness is synchronically mereologically determined by its lower-level physical realizers and each lower-level physical state is causally determined by an antecedent lower-level physical state, then there is nothing for consciousness as such to do in this world, on pain of denying physicalism, physics or some other 'absurdity'.

Notice that this argument has nothing to do with consciousness (or the mental) qua consciousness! The argument could just as well be about biological properties for example. Therefore refuting the argument leaves open the possibility of another argument aimed against the efficacy of consciousness as such. Oddly, such arguments are rare in the literature on mental causation in the last couple of decades.

This argument is generally aimed at British (classical) emergentism, non-reductive physicalism or any view that holds the mental is not a physical type but which accepts the realization principle and/or causal closure (Kim, 1999). If we construe causal closure as forbidding mental causes of physical effects and we hold that the mental is non-physical then that alone would entail epiphenomenalism. And again, epiphenomenalism is troubling because 'it just seems crazy — or, to put it more politely, seriously counterintuitive. Is it really a serious possibility that pains don't cause hands to withdraw from fires (and by virtue of being painful)? Do our thoughts not control our actions, what we say?' (Levine, 2001, p. 23). It is the absurdity of epiphenomenalism upon which reductive physicalism often rests its case: 'Thus only if mental phenomena are somehow constructible from, or constituted by [identifiable with], the physical phenomena that serve as the ultimate causal basis for all changes in the distribution of matter and energy does it seem possible to make sense out of mental–physical causal interaction' (Levine, 2001, p. 5).

The implication here is that if every conscious state is identifiable with its particular physical realizing state (such as a particular brain state), then the problem of mental causation has been solved. But even granting the identity of any given conscious state with a particular brain state, given the preceding argument, conscious states in virtue of being conscious will be epiphenomenal because by hypothesis, only the fundamental physical realizers are truly efficacious: 'There is no way around it. If materialism is true, then all causal efficacy is constituted ultimately by the basic physical properties ... then of course it will turn out that mental properties, along with all other non-basic physical properties, are not efficacious' (Levine, 2001, p. 28). Monism is a necessary but not sufficient condition for accounting for mental causation.

Physicalism (reductive and non-reductive) is in a bind regarding mental causation because it implies the epiphenomenalism of the conscious qua conscious if the realization principle is true *and* if it isn't! At this juncture reductive and non-reductive physicalism both give the same reply: 'Rather, what makes it [mental causation] a genuine case of causation is the fact that there is a lawful regularity that holds between beliefs and desires with certain contents and behaviors of the relevant kinds. True, there are lower-level physical mechanisms that sustain the regularity, but this doesn't itself take away from the regularity's status as a lawful regularity. It supports counterfactuals, is confirmed by instances, and, I believe, grounds singular causal claims' (Levine, 2001, pp. 28–9). After all, the physicalist will point out, we talk about all kinds of properties as being causal (such as hardness, liquidity, sharpness, etc.), but upon reflection we don't really believe such properties are causal so described. Hardness isn't efficacious qua hardness but rather in each case there is a more perspicuous story to tell about the molecular structures involved (and so on).

I think that this reply on the part of physicalism might seem reasonable until you reflect on its metaphysical commitments: namely that all the macroscopic (diachronic) regularities that we describe and explain in terms of historical processes, psychological processes, or other such macroscopic, self-contained coherent narratives, are all in reality determined (whether synchronically or diachronically) from the 'bottom-up' by fundamental physical forces and the laws that govern them. Why and how would the interplay of fundamental physical forces give rise to a world that is coherent and 'meaningful' from the perspective of these higher-level narratives? What are the odds of that happening? For example, how and why did fundamental physics 'conspire' to produce World War II on the heels of World War I and make that subsequent event seem so historically/sociologically obvious or likely as an historical/sociological event? Why does the world make so much sense from these higher-levels perspectives, why isn't the world more like a disconnected crazy-quilt of events than a coherent narrative or story? Think back to the Cambodian woman's Conversion Disorder driven blindness. The obvious causal account of this event is that she had a traumatic experience that, because of her love for her children, her guilt and shock subsequently induced 'hysterical' blindness in her. It would be one thing if physicalism merely wanted to assert that the physical causal mechanism of her blindness is neurochemical. However if physicalism is true then the ultimate causal account of her blindness is as follows: fundamental physical forces 'blindly' instantiated (determined from the 'bottom-up') a story line whereby her children were killed and she went blind.

Another counterintuitive aspect of physicalism's account of mental causation is the analogy of consciousness with properties like hardness. We can make sense of the notion that a diamond is realized by or identifiable with a certain molecular structure. We have some understanding of what it means to say that a Universal Turing Machine is realized by a particular computer. We can even make sense of the claim that the property of hardness is realized by certain (diverse) molecular structures. But what does it even mean to say that consciousness as consciousness is realized in or by a brain? The realization relation seems to imply some kind of part/whole relationship (though the whole may be functionally characterized) or constitutive relationship, but the relationship between consciousness as such and neural assemblies is not obviously such a relationship — indeed, this seems like a category mistake unless one assumes that consciousness as such is nothing but it's functional role.

Now obviously, the fact that physicalism is radically counterintuitive doesn't mean it's false, but why should we believe it? I contend that we should only believe something so counterintuitive and inexplicable relative to our first-person conscious experience and explanatory schemas if there is extremely strong evidence in its favour. But is there? Does the preceding argument constitute such grounds? The answer, I will argue, is 'no'. What motivates the realization principle, the causal closure of the physical domain and the causal inheritance principle? Extended arguments for these three claims are thin on the ground but the following are the most common motivations given in the literature: (1) The actuality/possibility of mental causation requires they be true (Levine 2001); (2) Mereological supervenience (Kim, 1999); (3) The 'layered' or hierarchical model of the world (Kim, 1998); (4) Physicalism requires they be true (Kim, 1993); (5) Physics requires they be true, e.g., the completeness of physics (Levine, 2001; McLaughlin, 1992); and (6) Causal determinism (Levine, 2001). I shall address each in turn.

We have just seen why (1) won't work as a justification: the argument, if true, makes mental causation impossible; (2) and (3) do justify the argument, but I have been arguing throughout that they are probably false. Mereological emergence just is the denial of both these points. Each mental property need not supervene on some physical property to come into existence-the relationship between so-called 'levels' (which in reality is a tangled web) is not one of realization, supervenience or determination. For example, in the case of entanglement, the subvenient 'physical' properties (assuming they ever existed in the first place as independent entities, a dubious assumption given quantum mechanics) get 'sucked up' (if you will) in producing the entangled state. In the case of irreducibly relational properties where the relations themselves are the most basic unit, the relata (intrinsic properties) lose their autonomy. By analogy, such compound mental properties could be involved in causal processes in which their subvenient physical property instances do not co-occur with the mental property instances (compounds). This alleviates worries about both over-determination and epiphenomenalism (Humphreys, 1997). This also allows for a diachronic and horizontal causal process of compound mental properties. Perhaps the compound properties throughout the web of the world 'wink' in and out of existence through a constant process of coherence and decoherence (to borrow a metaphor from quantum mechanics) in intricate and deeply interconnected ways. There is nothing truly 'downward' about such causal processes, as there are no vital forces involved, no strict top-to-bottom hierarchy of properties and no causal competition between emergent properties and their co-occurring 'realization base'. Therefore the causal inheritance principle never comes into play.

Causal closure of the physical domain is violated by mereological emergence if we take the claim to be about the lowest level microphysical domain (and not about the physical domain versus the non-physical/mental domain as such). However this violation of mereological causal closure (as it were) does not necessarily mean a violation of physical domain causal closure in general. Causal closure construed mereologically is trivially violated by mereological emergence because the lower-level property instances lose whatever individual existence they had in producing the resulting compound, emergent property. On the other interpretation, where causal closure refers now to the physical domain as opposed to the non-physical/mental domain, then causal closure is not violated if nomological supervenience is true and causal consciousness turns out to be a quantum property (i.e., a physical property). If on the other hand nomological emergence is true, and causal consciousness is not quantum in nature, then causal closure of the physical domain in general would be violated. But again, if the world is as entangled as mereological emergence suggests, then the very distinction between the physical and non-physical may be contextual.

Regarding justification (4), it is true that the essence of physicalism (as generally conceived) is embodied by the preceding argument and therefore, if the realization principle, causal closure, etc., are false then physicalism is false. But if I am right about the evidence from physics itself, then quantum mechanics forces one to choose between an *a priori* physicalism which has mereological supervenience as one of its first axioms and methodological/empirical physicalism (based on physics itself) which can reasonably accept mereological emergence based on the evidence and our best theories to date.

Contrary to what I have argued, it is alleged by many that physics itself justifies the argument. This is justification (5). This claim is usually based on the 'completeness of physics': physics alone is self-contained; its explanations need make no appeal to the laws of other sciences, and its laws cover the entities, properties and laws of the special sciences. Thus the completeness claim has two aspects, the autonomy of fundamental microphysics and the universality of fundamental microphysics. Take the following: 'Bodily movements involve changes in distribution of matter and energy. But from physics we know that the only forces that can affect such distributions are those realized in the fundamental physical properties' (Levine, 2001, p. 16; see also pp. 5, 22, 36 and 38). And even more forcefully: 'The lattice forces that hold together organic molecules are electromagnetic in origin. As truly remarkable as it is, it seems to be a fact about our world that the fundamental forces which influence acceleration are all exerted at the subatomic level' (McLaughlin, 1992, p. 91). But given everything I have said, it should be clear that such claims are tendentious and contentious. That is, scientifically informed people can reasonably disagree about such claims.

Unless the preceding passages are just restatements of mereological supervenience, that is assuming they don't simply beg the question, what follows from such claims is not the impossibility of mental causation but rather that mental causation requires that consciousness as such be able to influence the distribution of matter and energy via fundamental physical properties. But this is precisely what completeness forbids unless consciousness is itself a fundamental physical property: 'events like hand movements are physical events, covered by the laws of physics. We have no reason currently to believe that the physical trajectories of the parts of the hand are causally determined in any way differently from the physical trajectories of objects for which no mental causes are ever hypothesized' (Levine, 2001, p. 22).

While the assumption of the *in principle* completeness of physics is an essential axiom of *a priori physicalism*, how does it fare with *methodological/*

empirical physicalism, with physics in practice? Not very well as we have already seen. There are two primary reasons for the return of emergentism in general and for the 'disunity of science' movement in particular (Cartwright, 1999; Dupré, 1993), and we have already discussed both of these reasons in detail. First, there is the possibility of mereological emergence within fundamental physics itself. Second, a growing body of literature focusing on actual scientific practice suggests that there really are not many cases of successful intertheoretic reduction in the empiricist tradition of *demonstrating* nomological supervenience.

Thus, actual scientific practice not only calls into question mereological supervenience, but it calls into question nomological supervenience as well. In short, *methodological/empirical physicalism* calls into question the best grounds for believing in the completeness of physics in the first place! In practice physics is not complete. Nor at this juncture is it at all obvious that the incompleteness of physics is best explained by appeal to mere predictive/explanatory emergence. So when Levine says 'we have no reason currently to believe that the physical trajectories of the parts of the hand are causally determined in any way differently from the physical trajectories of [merely physical]objects', there are two things to keep in mind. First, we have our first-person conscious experience which tells us that, unlike merely physical objects, we are 'volitional' beings in virtue of being conscious. Second, based on actual scientific practice it is still an open empirical question as to whether or not chemical, macrophysical and biological phenomena mereologically and/or nomologically supervene on fundamental physical phenomena and laws, so how can it be regarded as a *scientific fact* that human behavior and mental phenomena so supervene. Of course *a priori physicalism* is free to reply that any present indications of either mereological or nomological emergence are just a function of ignorance, such indicators are really just signs of predictive/explanatory emergence. Only time will tell (maybe). But what I have argued is that consciousness as an emergent and causally efficacious feature of the world is at least as plausible as *a priori physicalism*. Indeed, I hope to have shown something even stronger. Namely, that given our scientific knowledge to date and given our first-person conscious experience, emergence is a *more* plausible and scientifically sober framework than *a priori physicalism*.

There is not much to say about justification (6): causal determinism. This is because it is still an open scientific question as to whether or not either the microphysical or the macroscopic domains are deterministic or stochastic (Earman, 1986; Humphreys, 1989; Bishop/Kronz, 1999). For a recent account of how probabilistic causation would make mental causation possible see Sober (1999). In short, a similar point applies in that scientifically informed people can reasonably disagree about the status of (macroscopic or microscopic) causal determinism.

Objection # 2: Emergence and the ontological/explanatory gap

This brings me to the second main objection: emergence cannot bridge the ontological/explanatory gap between matter and consciousness. This objection is the cornerstone of fundamental property dualism (Chalmers, 1996) and pan-experientialism (Griffin, 1998), because it is supposed to show that neither physicalism nor emergentism can even in principle account for consciousness. Take the following:

> The mental properties of the complex organism must result from some properties of its basic components, suitably combined: and these cannot be merely physical properties [meaning non-conscious?], or else in combination they will yield nothing but other physical properties' [meaning non-conscious?] (Nagel, 1986, p. 49).

> It is inconceivable that a subject, something that it is like something to be, could arise out of mere objects naturally. *This* type of alleged emergence violates the principle of continuity [no 'miracles' allowed] and thereby implies a violation of naturalism (Griffin, 1998, p. 64).

> We have here a prime example of a category mistake. The alleged emergence of subjectivity out of pure objectivity has been said to be analogous to examples of emergence that are different in kind. But the alleged emergence of experience is not simply one more example of such emergence. It involves instead the alleged emergence of an 'inside' from things that have only outsides (Griffin, 1998, pp. 64–5)

The problem with this objection against emergence, as the preceding passages make clear, is that it begs the question. The objection only goes through if we allow that 'physical' implies *essentially/inherently* non-conscious or non-mental. But barring a Cartesian worldview with a Cartesian view of matter, I see no reason to believe that any such implication holds for matter as conceived by modern physics. Fundamental property dualism and panexperientialism need to tell us *why exactly* consciousness cannot arise from that which is non-conscious. Chalmers (1996) does try to provide such an argument, but as Van Gulick notes, his argument rests on two key premises that have been called into question here and elsewhere (see Van Gulick, 2001 [this volume], pp. 11–13): (1) The theoretical-derivational account of intertheoretic reduction with *a priori* conceptual necessity, and (2) That everything in the world *except for* conscious states mereologically and nomologically supervenes on fundamental physical states. In short, Chalmers is just wrong when he asserts the following:

> Materialism is true, if all facts follow from physical facts. As I argue at length in my book, in *most* domains [with the exception of consciousness] it seems that they certainly do. The low-level facts about physical entities determine the facts about physical structure and function at all levels with conceptual necessity, which is enough to determine the facts about chemistry, about biology, and so on (Chalmers, 1997, p. 12).

Given these two assumptions, then emergence could not bridge the ontological/explanatory gap: the presence of consciousness could never be *scientifically explained* because consciousness could not exist in such a world unless it were a fundamental ingredient. However emergentism explicitly rejects both these assumptions. Consciousness is just one emergent feature of the world among many. The emergence of consciousness is certainly more counterintuitive and harder to explain scientifically (here at the dawn of the mind-sciences) than other cases of emergence, but it is not miraculous simply because we cannot currently explain it.

We must also remember that any monist who rejects full-blooded idealism must also grant that (full-blooded) consciousness did indeed *somehow* arise from that which is not conscious. As Van Gulick notes (2001 [this volume]), this is the source of a well-known dilemma for both fundamental dualism and panexperientialism:

> The idea of proto-psychic features also confronts a dilemma: the more we view them as *like* familiar mental properties the more implausible it is to suppose that they could be universally present in simple physical components — how could a molecule be aware, or how could an atom experience red or feel pain? But conversely the more we view them as *unlike* familiar mental features, i.e., the more we emphasize their 'proto-ness', the more difficult it is to see how they might give rise to consciousness. The basic explanatory gap just reoccurs at the boundary between the proto-psychic and the conscious (p. 24).

Significantly, most panexperientialists (of both the present and the past) have appealed exclusively to both mereological emergence and nomological emergence to bridge the explanatory/ontological gap between the proto-psychic and the conscious (see: Whitehead, 1978, pp. 45–6, 88, 187 and 229; Griffin, 1998, pp. 94, 168–9, 178–94, 233–40; Seager, 1995; 1999). And as I suggested at the beginning, it is unlikely that accepting either fundamental dualism or panexperientialism is going to provide any advantage whatsoever for discovering the *scientific explanation* for the emergence of consciousness. Just to be clear, my point here is not that I prefer physicalism over fundamental dualism or panexperientialism when it comes to an account of the fundamental ingredients of the world. Rather, barring full-blooded idealism, I see no *empirically useful* difference between any of these three views when it comes to the nature of the fundamental ingredients. I suspect that the fundamental element of the world of which both matter and mind are but aspects, is neither mental nor physical as we generally conceive these categories. My suspicion is based on our best accounts of quantum gravity (such as string theory and M-theory) as well as pregeometric accounts of fundamental reality (Callender and Huggett, 2001).

I think that emergence is also scientifically more credible than physicalism when it comes to bridging the ontological/explanatory gap. The three most plausible routes for physicalism to bridge the ontological/explanatory gap are as follows:

(1) Physicalism can hold that there are fundamental/basic (bottom level) psycho-physical bridge laws that explain the correlation between conscious states and physical states. Given the transitivity of realization relations, this is what physicalism is probably committed to. Indeed, physicalism is defined by mereological and nomological supervenience which imply that psycho-physical bridge-laws must ultimately be basic, because all the brute facts and brute laws are at the fundamental level of reality on this view. Chalmers, though not a physicalist, advocates just this view: 'Once we introduce fundamental psychophysical laws into our picture of nature, the explanatory gap has itself been explained: it is only

to be expected, given that nature is the way it is' (1997, p. 19). But the idea of fundamental (brute) laws that relate conscious states and physical states seems counterintuitive as well as antithetical to physicalism. Does physicalism really expect, for example, that superstring theory will have laws relating such states? Chalmers at least seems to appreciate the problem: 'It would be very odd if there were *fundamental* laws connecting complex organizations to experience' (1997, p. 32). So it would appear that physicalism is committed to something that is troubling and counterintuitive, especially from the point of view of physicalism itself.

(2) Physicalism can hold that psycho-physical bridge laws are brute *macroscopic* laws. But this is a violation of nomological supervenience, a major tenant of physicalism. Such laws would be 'nomological danglers' from the perspective of physicalism.

So it seems that the notion of brute psycho-physical laws, whether they be *fundamental* or *macroscopic/higher-level* laws, is troubling for physicalism. I think this is because in a world in which physicalism is true, there just shouldn't be any 'conscious remainders' at any level. Recall that it was the brute psycho-physical bridge laws of British emergentism that made it the object of such derision on the part of reductive physicalism (McLaughlin, 1992). This brings us to:

(3) Physicalism can hold that 'because [according to reductive physicalism] it is an identity that is involved there is no explanatory gap to bridge. There are not two things whose linkage needs to be explained; there is just one thing, and it like everything else is necessarily identical with itself and not with anything else. If Brian's pain just *is* a certain pattern of brain activity in the *identity* sense of "is", then there is no gap to be closed any more than there is in any case of identity' (Van Gulick, 2001 [this volume], p. 6).

But the case of mind/brain identity does not seem to be like other cases of identity: 'With the standard cases [of identity], once all the relevant empirical information is supplied, any request for explanation of the identities is quite unintelligible, not in this [mind/body] case!' (Levine, 2001, p. 92). For example, if someone asks how H_2O could possibly possess the properties deemed essential to water (such as its liquidity) we can answer their question with an explanation from quantum chemistry. But we cannot answer the analogous question about consciousness and the brain. Accepting the identity theory as a *philosophical* solution to the mind–body problem does not obviate the need for a *scientific explanation* of how brain states can have consciousness as one of their properties; just like the case of water and liquidity. As Levine puts it, 'if materialism is true, there ought to be an explanation of how the mental arises from the physical: a realization theory . . . such a theory should in principle be accessible' (2001, p. 69). Chalmers makes exactly the same point: 'in postulating an explanatory "identity", one is trying to get something for nothing: all of the explanatory work of a fundamental law, at none of the ontological cost' (1997, p. 12). Van Gulick also makes the same point in this issue. Furthermore, while the identity theory seems plausible when we talk about identifying conscious *states* with brain *states*, it seems less plausible when we consider that each brain state has conscious *aspects* such as feeling pain and non-conscious *aspects* such as c-fibre activation. How can these two disparate *aspects* of a particular brain state be *identical* without risking eliminativism?

VI: Concluding Remarks —
Wilder Speculations Still

Regarding the ultimate fate of mereological and nomological emergence respectively, there are two general possibilities. Either these respective forms of emergence are merely a function of our ignorance or they are real facts about the world. If they are real facts about the world then they may be either universally true or restricted to a particular domain such as microphysics. Of course the ultimate fate of mereological emergence might be different from that of nomological emergence and vice-versa. Philosophers and physicists wedded to ancient atomism will no doubt opt for the ignorance interpretation. Such an interpretation will be harder to maintain for mereological emergence as there is good general empirical and specific experimental evidence for quantum holism (Hawthorne and Silberstein, 1995).

We have seen however, regarding the ignorance interpretation, that the facts to date (actual physics itself and actual scientific practice) do not in all probability support *a priori physicalism*. Contrary to the hype, based on the evidence alone, it is much too early in the game to decide against mereological and nomological emergence. The best explanation for the pervasive predictive/explanatory emergence within the natural sciences and beyond, may be nomological emergence. If one takes quantum holism seriously and the notion of intertheoretic relations as singular asymptotic expansions seriously, then the various so-called domains of the world (physical, chemical, biological, mental and those in between) do not appear to form a discrete hierarchy. Rather these domain relations look more like non-Boolean lattices; the various domains will have overlapping areas or unions, but they will not be co-extensional (Primas, 1991). So physical properties in one domain may be necessary for properties in another domain to emerge, but not sufficient. Likewise, physical phenomena may be necessary but not sufficient for consciousness. If quantum holism is universally valid then the ancient atomistic idea of a world constructed out of individual and independent parts (such as atoms) is bankrupt. Given universal nonseparability (even including space-like separated events), there are no individual and independent parts; the universe is truly an undivided whole. Symmetry breaking suggests a similar picture.

On the other hand, consciousness might be *literally* quantum mechanical or fundamentally physical in nature. That, I take it, is an open empirical question. How can we resolve this question? If consciousness is not nomologically emergent with respect to quantum mechanics (or whatever the fundamental theory turns out to be) then it is reasonable to say that it is quantum mechanical in nature. That is, if it can be shown that quantum mechanics can *in principle* derive or otherwise 'explain from below' consciousness, then consciousness is continuous with quantum mechanics. If however consciousness is nomologically emergent with respect to quantum mechanics, then consciousness is not quantum mechanical in nature. Notice that neither possible outcome threatens either the reality or the efficacy of consciousness.

I think that mereological and (probably) nomological emergence are at least necessary conditions for the naturalistic explanation of consciousness. If the universe is as conceived by ancient atomism, if classical mechanics is a complete picture of the world, then it becomes impossible to account for consciousness — to really *explain* consciousness. The irony here is that if one wants to maintain physicalism characterized as the view that the fundamental ingredients of the world are non-conscious, then one has to give up physicalism as characterized by mereological and nomological supervenience, and vice-versa. Something has got to give. As Van Gulick notes, I believe we need conceptual revolutions with respect to both our notions of consciousness and matter before we can explain consciousness. However, perhaps physics itself already contains the seeds of such a revolution.

I do doubt however that a 'final theory' of consciousness can be as simple as some single neurochemical or quantum mechanical mechanism. First, there may be no shared mechanism (such as a neurochemical feature) that explains why disparate mental states (such as the various sensory and memory modalities) are conscious states; assuming for the moment that it makes sense to talk of such states as being conscious in isolation from the system as a whole. That is, there may be multiple 'mechanisms' of emergence for basic states of awareness such as sensory experience. Based on everything we know about the continuum of animal and human neurophysiology and their various conscious states, this seems probable. Second, if consciousness is like superposition/entangled states or even a dynamical property, then it might not even be localizable in *physical space* (such as the brain) in the way suggested by the search for the neural correlates of consciousness (NCC), rather, higher-states of consciousness (such as self-awareness and explicit cognition) might be a mereologically emergent, global and systemic property of the *brain-world complex as a whole*. This is speculative to be sure. However it seems likely that many higher-states of consciousness such as self-consciousness and symbol processing are continuous with the socio-cultural domain. For example, an individual is unlikely to develop such higher-states in (severe) social isolation. In short, explaining *unity and subjectivity* (at least in humans) might require more than mechanisms 'inside the head.' Maybe some features of consciousness are best seen as *not in the head per se*, but distributed over complex interactions between many individuals across time. Third, an even more troubling possibility from the perspective of ancient atomism: perhaps there are no truly intrinsic properties at all. Maybe the atomistic notion of fundamental intrinsicality at bottom is nothing more than a guiding myth of science.

In addition to encouraging us to look for mechanisms of emergence, rather than thinking of consciousness as literally quantum mechanical, I think quantum mechanics and its founding fathers provides us with a potentially powerful analogy regarding the relationship between brain states and consciousness. Perhaps mind (mental variables/observables) and matter (physical variables/ observables) are *nonseparable* (or even related via *complementarity*) aspects that emerge from something that is neither mental nor physical *per se*. Something like this view is suggested by the writings of more than one founding father of quantum mechanics (Bohr, 1948; Pauli, 1948; Wigner, 1970). The suggestion here is that even if consciousness is not literally quantum mechanical, quantum mechanics though technically only a theory of the microcosm, is a universal theory such that some of its insights and relations can be mapped onto the macrocosm. This is an ancient metaphysical idea to be sure.

Nonseparability we have already discussed, so what about this notion of complementarity? Properties related by complementarity in quantum mechanics are non-commuting properties that are mutually exclusive and complement each other. Such properties are essential aspects of the same entity or system. Perhaps the most well known formal example is the non-commuting relationship between position and momentum. Perhaps the most well known (more) informal example is wave/particle duality. There is a relationship between nonseparability and complementarity at least in the sense that the non-commutativity of observables in quantum mechanics (the non-Boolean structure of the theory) accounts for the latter but is also a necessary condition for the former. Non-commutative observables are sufficient for uncertainty relations (such as Heisenberg's uncertainty) and (at least) necessary for entangled states.

It seems to me that this analogy of complementarity/nonseparability is highly suggestive of the relationship between mind and matter in a number of ways: (1) The relationship is between properties of a single entity/process; (2) The properties involved (subjective versus objective) and/or their best descriptions are mutually exclusive of one another and therefore cannot be identified or reduced to one another in the *nothing but* sense of these relations; (3) The properties involved are highly correlated statistically and phenomenologically, so as to be easily confused with efficient causal relations or identity relations; (4) The properties involved each represent mutually exclusive but complementary aspects of one thing that each have (by definition) unique causal capacities in their own right, nor must we posit efficient causal relations between such properties to explain their efficacy; and (5) Such properties are not higher-level and lower-level properties related hierarchically or via realization/constitution, they are properties characterizable as being on 'the same level' as it were. These five points go a long way toward simultaneously capturing the intuitions of identity accounts, causal accounts and emergentist accounts of the mind/body relationship. I see no reason why the 'mental' cannot be usefully viewed as phenomenal *and* physical (or biological) *and* functional. Obviously, making this analogy more rigorous would require spelling out a number of details such as exactly what properties are being alleged to be nonseparable from one another, etc. (For more recent proposals in the spirit of this view about the relationship between matter and consciousness see Atmanspacher, forthcoming; Walach & Römer, 2000).

Leaving aside the viability of the preceding analogy from quantum mechanics for the moment, if we take seriously the idea that consciousness is an emergent phenomenon, there is another way in which quantum mechanics and fundamental physics can be of use in explaining consciousness. We can examine the class of emergent phenomena in physics to see what crucial properties they have in common. For example, it turns out that a number of both mereologically and (apparently) nomologically emergent phenomena can ultimately be best explained in terms of *symmetry breaking* in one related form or another (Anderson, 1994; Silberstein, 1998; Atmanspacher, forthcoming). This is true, for example of all the cases discussed in this paper including nonseparability/nonlocality (Silberstein, 1998) and the various types of phase transitions from condensed matter theory and the standard model (Anderson, 1994). For corresponding applications beyond conventional physics see Atmanspacher *et al.* (2001). Getting clear on the key universal aspects of predicting and explaining emergence, whether it be symmetry breaking of some sort, singular asymptotic expansions or what have you, might help us see how to extend such patterns of explanation to the case of consciousness. Unfortunately, the disanalogy here is that in other cases of attempted intertheortic reduction we have two fairly well worked out theories, such as thermodynamics and statistical mechanics. But the case of consciousness is different. Neuroscience does not offer a mature theory and psychology (folk, cognitive or otherwise) may not lead to a (scientific) theory at all, mature or not. Given the long road ahead, I counsel agnosticism and patience — the true spirit of science. In the meantime, I hope to have provided a plausible set of background assumptions for a unified research programme.

References

Anderson, P.W. (1994), *A Career in Theoretical Physics* (Singapore: World Scientific Publishing).

Antony, L. (1999), 'Making room for the mental: Comments on Kim's "Making Sense of Emergence"', *Philosophical Studies*, **95** (2), pp. 37–43.

Antony, L. & Levine, J. (1997), 'Reduction with autonomy', in *Philosophical Perspectives 11, Mind, Causation, and World* (Boston, MA: Blackwell).

Atmanspacher, H. (forthcoming), 'Mind and matter as asymptotically disjoint, inequivalent representations with broken time-reversal symmetry' (under review).

Atmanspacher, H., Römer, H. and Walach, H. (2001), 'Weak quantum theory: Complementarity and entanglement in physics and beyond' (to be published in the proceedings of ECHO 4).

Auyang, S. (1998), *Foundations of Complex-System Theories* (Cambridge: Cambridge University Press).

Batterman, R.W. (2000), 'Multiple realizability and universality', *British Journal of the Philosophy of Science*, 51, pp. 115–45.

Batterman, R. (2001), *The Devil in the Details: Asymptotic Reasoning in Explanation, Reduction, and Emergence* (Oxford: Oxford University Press).

Beckermann, A., Flohr, H. & Kim, J. (ed. 1992), *Emergence or Reduction? Essays on the Prospects for Nonreductive Physicalism* (Berlin: DeGruyter).

Bedau, M. (1997) 'Weak emergence', in *Philosophical Perspectives (11): Mind, Causation, and World*, ed. J.E. Tomberlin (Boston, MA: Blackwell).

Belot, G. & Earman, J. (1997), 'Chaos out of order: Quantum mechanics, the correspondence principle and chaos', *Studies in History and Philosophy of Modern Physics*, **2**, pp. 147–82.

Berry, M.V. (1994), 'Singularities in waves and rays' in *Physics of Defects (Les Houches, Session XXXV, 1980)*, ed. R. Balian, M. Kléman and J.P. Poirier (Amsterdam: North Holland).

Berry, M.V. (2000), 'Chaos and the semiclassical limit of quantum mechanics (is the moon there when somebody looks?)', *Proceedings for the Vatican Conference on Quantum Mechanics and Quantum Field Theory*, pp. 2–26.

Berry, M.V. and Howls, C. (1993), 'Infinity interpreted', *Physics World*, **June**, pp. 35–9.

Bickle, J. (1998), *Psychoneuronal Reduction: The New Wave* (Cambridge, MA:MIT Press).

Bishop, R. & Kronz, F. (1999), 'Is chaos indeterministic?', in *Language, Quantum, Music*, ed. M.L. Chiara *et al.* (Dordrecht: Kluwer).
Block, N. (1995), 'On a confusion about a function of consciousness', *Behavioral and Brain Sciences*, **18** (2), pp. 227–87.
Bohm, D. & Hiley, B.J. (1993), *The Undivided Universe* (London: Routledge).
Bohr, N. (1934), *Atomic Theory and the Descriptive of Nature* (Cambridge: Cambridge University Press).
Bohr, N. (1948), 'On the notions of causality and complementarity', *Dialectica*, **2**, pp. 312–19.
Cartwright, N. (1999), *The Dappled World: A Study of the Boundaries of Science* (New York: Cambridge University Press).
Cartwright, N. (1997), 'Why physics?' in *The Large, the Small and the Human Mind*, ed. Penrose, R., Shimony, A., Cartwright, N. & Hawking, S. (Cambridge: Cambridge University Press).
Causey, R.L. (1977), *Unity of Science* (Dodrecht: Reidel).
Callender, C. and Huggett, N. (2001), *Physics Meets Philosophy at the Planck Scale* (Cambridge: Cambridge University Press).
Chalmers, D.J. (1996), *The Conscious Mind* (New York: Oxford University Press).
Chalmers, D.J. (1997), 'Moving forward on the problem of consciousness', *Journal of Consciousness Studies*, **4** (1), pp. 3–46.
Churchland, P.M. (1989), *A Neurocomputational Perspective: The Nature of Mind and the Structure of Science* (Cambridge, MA: MIT Press).
Churchland, P.S. (1986), *Neurophilosophy: Toward a Unified Science of the Mind-Brain* (Cambridge, MA: MIT Press).
Cornell, E. A. & Wieman, C.E. (1998), 'The Bose-Einstein condensate' *Scientific American*, **278** (3), pp. 40–5.
Crutchfield, J. (1994), 'Is anything ever new? Considering emergence', *Complexity: Metaphors, Models, and Reality*, ed. Cowan, G. *et al.* (Cambridge, MA: Perseus Publishing).
Dupré, J. (1993), *The Disorder of Things: Metaphysical Foundations of the Disunity of Science* (Boston, MA: Harvard University Press).
Earman, J. (1986), *A Primer on Determinism* (Dordrecht: Reidel).
Feyerabend, P.K. (1962), 'Explanation, reduction and empiricism' in *Minnesota Studies in the Philosophy of Science III*, ed. Feigl & Maxwell (Minnesota: Universe of Minnesota Press).
Flanagan, O. (1992), *Consciousness Reconsidered* (Cambridge, MA: MIT Press).
Giere, R. (1988), *Explaining Science: A Cognitive Approach* (Chicago, IL: University of Chicago Press).
Grantham, T. and Nichols, S. (2000), 'Adaptive complexity and phenomenal consciousness', *Philosophy of Science*, **67** (4), pp. 648–70.
Griffin, D.R. (1998), *Unsnarling the World-Knot: Consciousness, Freedom and the Mind-Body Problem* (University of California Press).
Hardcastle, V. (1999), *The Myth of Pain* (Boston, MA: MIT Press).
Hawthorne, J. & Silberstein, M. (1995), 'For whom the Bell arguments toll' *Synthese*, **102**, pp. 99–138.
Healey, R.A. (1991), 'Holism and nonseparability', *Journal of Philosophy*, **88**, pp. 393–421.
Hellman, G. (1999) 'Reduction (?) to what (?): Comments on L. Sklar's "The reduction (?) of thermodynamics to statistical mechanics"', *Philosophical Studies*, **95** (1–2), pp. 200–13.
Hendry, R (1998), 'Models and approximations in quantum chemistry', in *Idealization in Contemporary Physics*, ed. N. Shanks (Amsterdam: Rodopi).
Hendry, R. (1999) 'Molecular models and the question of physicalism', *Hyle*, **5** (2), pp. 143–60.
Hooker, C.A. (1981), 'Towards a General Theory of Reduction. Part I: Historical and Scientific Setting. Part II: Identity in Reduction. Part III: Cross-Categorical Reduction', *Dialogue*, **20**, pp. 38–59, 201–36, 496–529.
Horst, S. (1999), 'Evolutionary explanation and the hard problem of consciousness', *Journal of Consciousness Studies*, **6** (1), pp. 39–49.
Humphreys, P. (1989), *The Chances of Explanation* (Princeton, NJ: Princeton University Press).
Humphreys, P. (1997), 'How properties emerge', *Philosophy of Science*, **64**, pp.1–17.
Kaku, M. (1993), *Quantum Field Theory: A Modern Introduction* (Oxford: Oxford University Press).
Kellert, S. (1993), *In the Wake of Chaos* (Chicago, IL: University of Chicago Press).
Kemeny, J. & Oppenheim, P. (1956), 'On reduction', *Philosophical Studies*, **7**, pp. 6–17.

Kim, J. (1993), 'Multiple realization and the metaphysics of reduction', in his *Supervenience and Mind* (Cambridge: Cambridge University Press).
Kim, J. (1996), *Philosophy of Mind* (Boulder, CO: Westview Press).
Kim, J. (1998), *Mind in a Physical World: An Essay on the Mind-Body Problem and Mental Causation* (Cambridge, MA: MIT Press).
Kim, J. (1999), 'Making sense of emergence', *Philosophical Studies*, **95** (1–2), pp. 3–36.
Kitcher, P. (1989), 'Explanatory unification and the causal structure of the world', in *Minnesota Studies in the Philosophy of Science, Volume 13, Scientific Explanation*, ed. P. Kitcher and W.C. Salmon (Minneapolis: University of Minneapolis Press).
Levine, J. (2001), *Purple Haze: The Puzzle of Consciousness* (New York: Oxford University Press).
Lewis, D. (1986), 'Causation', in his *Philosophical Papers, Volume II* (Oxford: Oxford University Press).
McGinn, C. (1999), *The Mysterious Flame: Conscious Minds in a Material World* (New York: Basic Books).
McLaughlin, B. (1992), 'Rise and fall of British emergentism', in Beckermann *et al.* (1992).
Machamer, P., Darden, L. & Craver, C.F. (2000), 'Thinking about mechanisms', *Philosophy of Science*, **67**, pp. 1–25.
Maudlin, T. (1998), 'Part and whole in quantum mechanics' in *Interpreting Bodies: Classical and Quantum Objects in Modern Physics*, ed. E. Castellani (Princeton, NJ: Princeton University Press).
Nagel, E. (1961), *The Structure of Science* (New York: Harcourt, Brace).
Nagel, T. (1986), *The View from Nowhere* (Oxford: Oxford University Press).
Oppenheim, P. & Putnam, H. (1958), 'Unity of science as a working hypothesis', in *Minnesota Studies in the Philosophy of Science*, ed. H. Feigl and G. Maxwell, *Philosophical Studies*, Vol. 95, pp. 45–90.
Pauli, W. (1948), 'Special issue on complementarity', *Dialectica*, **2**, pp. 300–11.
Primas, H. (1983), *Chemistry, Quantum Mechanics, and Reductionism* (Berlin: Springer-Verlag).
Primas, H. (1991), 'Reductionism: Palaver without precedent' in Agazzi (1991).
Primas, H. (1998), 'Emergence in the exact sciences', *Acta Polytechnica Scandinavica*, 91, pp. 83–98.
Ramsey, J.L. (1997), 'Molecular shape, reduction, explanation and approximate concepts', *Synthese*, **111**, pp. 233–51.
Revonsuo, A. (2001), 'Can functional brain imaging discover consciousness in the brain?' *Journal of Consciousness Studies*, **8** (3), pp. 3–23.
Scerri, E. (1994), 'Has chemistry been at least approximately reduced to quantum mechanics?', in *PSA 1994*, vol. 1., ed. D. Hull, M. Forbes & R. Burian (East Lansing, MI: Philosophy of Science Association), 160–70.
Schaffner, K.F. (1992), 'Philosophy of medicine' in *Introduction to the Philosophy of Science*, ed. M.H. Salmon *et al.* (Englewood Cliffs, NJ: Prentice Hall).
Schaffner, K.F. (1998), 'Genes, behavior, and developmental emergentism: One process, indivisible?', *Philosophy of Science*, **65**, pp. 209–52.
Seager, W. (1995), 'Consciousness, information and panpsychism', *Journal of Consciousness Studies*, **2** (3), pp. 272–88.
Seager, W. (1999), *Theories of Consciousness: An Introduction and Assessment* (New York: Routledge).
Seife, C. (2000), 'Cold numbers unmake the quantum mind', *Science*, **287**, 4 Feb. 2000.
Silberstein, M. (1998) 'Emergence and the mind/body problem', *Journal of Consciousness Studies*, **5** (4), pp. 464–82.
Silberstein, M. & McGeever, J. (1999), 'The search for ontological emergence', *The Philosophical Quarterly*, **49**, pp. 182–200.
Silberstein, M. and Machamer, P. (ed. 2002), *Blackwell Guide to the Philosophy of Science* (forthcoming).
Sklar, L. (1993), *Physics and Chance: Philosophical Issues in the Foundations of Statistical Mechanics* (Cambridge: Cambridge University Press).
Sklar, L. (1999), 'The reduction (?) of thermodynamics to statistical mechanics', *Philosophical Studies*, **95** (1–2), pp. 187–99.
Sober, E. (1999), 'Probabilistic causation', *Philosophical Studies*, **95** (1–2), August, pp. 44–60.
Stephan, A. (1992), 'Emergence: A systematic view on its historical facets' in Beckermann *et al.* (1992).

Tegmark, M. (2000), 'The quantum brain', *Physical Review E.* pp. 1–14.
Van Fraassen, B.C. (1989), *Laws and Symmetry* (Oxford: Oxford University Press).
Van Gulick, R. (2001), 'Reduction, emergence and other recent options on the mind/body problem: A philosophical overview', *Journal of Consciousness Studies*, **8** (9–10), pp. 1–34.
Walach, H. & Römer, H. (2000), 'Complementarity is a useful concept for consciousness', *Neuroendocrinology Letters*, **21**, pp. 221–32.
Whitehead, A.N. (1978), *Process and Reality* (New York: The Free Press).
Wigner, E. (1970), 'Physics and the explanation of life', *Found. Phys.*, **1**, pp. 35–45.
Wimsatt, W.C. (1976), 'Reductive explanation: A functional account,' in *Philosophy of Science Association*, ed. R.S. Cohen, C.A. Hooker, A.C. Michalos and van Evra (Dodrecht: Reidel).
Woody, A.I. (2000), 'Putting quantum mechanics to work in chemistry: The power of diagrammatic representation', *Philosophy of Science*: supplement to **67** (3), ed. D.A. Howard, Part II: Symposia Papers, pp. S612–27.

Scott Hagan and Masayuki Hirafuji

Constraints on an Emergent Formulation of Conscious Mental States

Fundamental limitations constraining the application of emergence to formulations of conscious mental states are explored within the paradigm of classical science. This paradigm includes standard interpretations of functionalism, computationalism and complex systems theories of mind — theories which are ultimately justified by an appeal to emergentist principles. We define a distinction between extrinsic and intrinsic accounts of emergent conscious states, and examine the prospects for both. Extrinsic accounts are subject to relativities with respect to external observers that must be resolved if the ontological character of conscious states is to be preserved. While this can, in some cases, be accomplished by imposing an appropriate invariance, no such strategy exists in the case of relativity with respect to boundary without absurd consequences. If, on the other hand, conscious states require intrinsic definition, a specification of the system boundary must be explicitly available if the conscious ontology is to be uniquely specified. Even minimal information requirements make this incompatible with locality constraints. We investigate what progress can be made in overcoming these obstacles by relaxing various assumptions.

Introduction

Emergence has recently become a popular strategy by which to protect the subject matter of the biological sciences and psychology from the ultimately reductive approach of physics (see for instance Scott, 1995, and references therein). Advocates suggest that emergent systems might be freed from epiphenomenal enslavement to the physical laws governing their most fundamental components, and thereby gain a separate ontological identity. One phenomenon in particular, that of consciousness, begs to have its ontology sharply distinguished from that of its physical substrate. It is particularly clear in this case that straightforward reduction leaves no room for conscious mental phenomena at all, making it an ideal candidate for the application of emergentist principles.

Although the concept of emergence is explicitly invoked only in the context of complex systems approaches to consciousness, it also lies – implicitly – at the heart of both functional and computational accounts. Here emergence provides the necessary principled distinction between functions and computations that are to be regarded as conscious and those that are not. Where this distinction is taken to be a matter of cognitive sophistication, it is by way of emergent criteria that each of these theories of mind indicates what physical systems suffice to bring into existence a conscious mental state. It is in terms of emergence that these models answer the critical dynamical questions that determine:

- when does the mechanism of consciousness become operable?
- what information exactly should be encompassed by the conscious state?
- why should mind be a necessary consequence of brain?

and so on. Other criteria have certainly been proposed to respond to these questions. Nevertheless, emergence often appears to underpin these proposals. Speed of computation has, for example, occasionally been invoked as a determining factor (P.S. Churchland, 1995). This, however, requires the prior determination of an emergent computational system and so does not constitute an independent criterion.

The fact that most mainstream positions in consciousness studies maintain an unarticulated reliance on a notion of emergence underscores the importance of critically examining the concept. We will not attempt to clarify the underlying physical assumptions of emergent formulation in general, but specifically with respect to the phenomenon of consciousness. Our discussion will be broad enough to encompass, for example, models of consciousness based on emergent properties like 40 Hz oscillations (Crick and Koch, 1990), neuronal gestalts (Greenfield, 1995), attractors (Skarda and Freeman, 1990; Freeman and Skarda, 1991; Port and Van Gelder, 1995) and vector activations (P.M. Churchland, 1995).

We undertake here to clarify whether or not the aims of emergent accounts of conscious mental states lie within the scope of a purely *classical* science. Since the mainstream neurobiological and cognitive science models of consciousness are all classical — that is, they make no specific appeal to quantum levels of description — a classical framework seems the appropriate context in which to investigate the use of emergence. This, of course, does not imply that these models reject quantum theory, but rather that they assume the explanatory power relevant to consciousness need not be sought at a quantum level.

It will be our contention that a plausible account of consciousness may demand more radical departures from the classical tradition in science than current theses suggest. We will also attempt to pinpoint what strategies might hold promise of circumventing the problems that beset many widely accepted theories.

Emergence in the Formulation of Conscious Mental States

There is no standard account of emergence that finds wide agreement (see, for instance Klee, 1984; Emmeche *et al.*, 1997; Collier, 1998). In order that our arguments might be applied to the broadest possible class of emergentist theories, therefore, no restrictive definition will be adopted. We must nevertheless characterize emergence. In this regard, certain criteria are invariably cited in the literature, most notably *novelty*. This specifies that emergence should engender novel features or properties at the level of the whole (so-called 'higher-order' features or properties) that are irreducible to features or properties of the primitive physical determinants. This will form the basis of our operative definition. In order for this criterion to be relevant with respect to consciousness, though, novelty must be given a specifically ontological reading. The emergence of conscious states cannot be merely epistemological without, at minimum, reducing consciousness to an epiphenomenal status.[1] While it is possible that *what* we think we are conscious of might be a consequence of the way we structure our concepts, it does not seem possible that the fact *that* we are conscious is equally susceptible to such an analysis. This is just the conclusion of the Cartesian *cogito*. The class of theories considered must therefore, at the outset, be narrowed to those in which conscious phenomena are granted an appropriate ontological status.

This study is interested in how conscious mental states can possibly be generated in the context of physicalism. To begin our analysis, we make a distinction between physicalist theories that are ultimately reductive and those that are ultimately emergentist. On the reductive side, straightforward identity theories encounter an immediate category error: there is no criterion on the basis of which to distinguish conscious states from physical states. Contemporary reductive approaches to consciousness generally skirt category error by explaining *away*[2] the phenomenon (see, for instance, Dennett, 1991) rather than allowing room for a mental ontology that is a natural and necessary consequence of the underlying dynamics. Emergent theories of consciousness as discussed here will be defined, in contrast to these kinds of straightforward reduction, such that a distinction of 'higher-order' stands between the physical and the conscious, and the conscious thereby preserves a novel ontology.

Proceeding on the assumption of emergentism, classical ontologies must either be adequate to an emergent account of consciousness or not. If classical ontologies are to be adequate, then emergence should at least be *compatible* with microphysical determinism. Microphysical determinism is, after all, the manner in which causal accounts are ultimately cashed out in a classical framework. Theories phrased in terms of complex systems easily accommodate such compatibility. These theories are in principle deterministic even though, in practice, they may become entirely unpredictable beyond a certain time frame. Mainstream functionalist and computationalist accounts are likewise compatible with

---

[1] One of the criteria by which to identify emergence — the criterion of unpredictability cited by complex systems accounts — must thus be considered specious when used to distinguish emergent conscious states. Unpredictability is a relational feature between observer and observed rather than a property inhering in the observed phenomenon itself, and hence has an epistemological rather than an ontological character.

[2] This is facilitated by a conflation of different senses of words like 'attention' and 'awareness' that permits purely physical treatments of the functional capacities to supplant a treatment of their specifically conscious aspects (see Chalmers, 1996, Ch. 1, for a discussion of this).

microphysical determinism. Note that *compatibility* does not require that the emergent story should be *reducible* to an account deterministically framed at the microphysical level. Rather, it requires only that there be no *contradiction* between the two. Nevertheless, this compels one to concede the closure of material-efficient causes.[3]

What is here excluded from consideration are theories that depart from microphysical determinism, as do for instance theories framed in terms of 'downward causation' (Campbell, 1974; see also Sperry, 1980; 1992). In such cases, the issue becomes one of determining the sufficiency conditions that must be satisfied in order to allow formal-final influences into the causal chain. Ali *et al.* (1998) characterize Sperry, for example, as following Polanyi (1967) in 'defining the formal-final causality of mind in field-theoretic terms, viz. as an autonomous boundary condition eliminating degrees of freedom in the lower level substrate'. On this view one *accepts* the sufficiency of mind to exert formal-final influence rather than defining conditions on which it is *determined* to be sufficient. In other words, there is no account of which, when and how physical systems become candidates to exert a formal-final influence on their physical constituents. Until such an account is actually spelled out, these non-standard conceptions of causation cannot be critically addressed in this framework.

Most mainstream interpretations of quantum theory are also not compatible with determinism at the micro-level and likewise lie beyond the scope of the current analysis. The notions of causation employed in interpretations of quantum theory constitute only a heuristic calculus and are not generally interpreted ontologically.[4]

Local and Global Bases

Restricting our attention to emergent theories formulated classically, consistent with standard accounts of causation, it is possible to further constrain the scope of inquiry by considering whether the physical substrate should consist in *local* or in *global* states.[5] 'Global' is used here in the sense of physics, where it is

---

[3] If, on the other hand, the closure of material-efficient causes is rejected, then one has already moved outside the context of classical science considered here. Particularly with regard to theories of consciousness coming from the natural sciences, it is presumed that such closure is generally implicit and intended.

[4] Bohm and Hiley (1993) elaborate an interpretation of quantum theory in which the particulate dynamics might be interpreted as being consistent with microphysical determinism. The treatment given is, moreover, explicitly causal and ontological. There is, however, an explicit violation of the classical locality principle that removes their interpretation from the scope of this paper.

[5] Note that this follows the traditional development of science in presupposing that the account should assume a *state* basis. Some process philosophers might regard this as fundamentally flawed, repudiating the interpretation of process, implied in any classical treatment, as the limit of a sequence of successive states. A novel interpretation of process, however, comes at a high price that necessarily involves the abandonment of much of the machinery of classical science. Even under this kind of radical re-orientation, much of the current critique, appropriately rephrased, might still apply. While an understanding of process may well contribute to the formulation of a theory of consciousness, the adoption of process philosophy (Whitehead, 1929/1985) does not, of itself, clearly solve the problems to be addressed in this article.

defined in contradistinction to 'local'. A global basis will be one for which the causally relevant substrate for an instantaneous conscious mental state is extended in space (or is extended over a space–time hyperslice, given relativistic description).

In a *purely* classical ontology — one that is classical 'all the way down' — local states could not be granted any physical extension at all. The physical basis for a local state could then only be a discrete point in space (or, phrased more relativistically, a discrete space–time event). Note here that the instantiation in something like a 'billiard ball' model of molecules artificially truncates structure below a certain level. The cohesion of the 'billiard ball' itself, for instance, and signals that propagate internal to its structure are irrelevant to explanations framed in terms of the model. Strictly speaking, such examples are 'local' only at scales larger than the one characterizing the 'billiard ball'. Here we will allow a model to count as 'local' if it is accounted for in terms of the most fundamental entities in the given classical ontology.

It is generally accepted that, at very small scales, classical science breaks down and it becomes necessary to invoke quantum theory. This is, of course, not to repudiate a classical basis for consciousness. Quantum theory need not enter, in this context, in any more than a trivial sense, leaving the explanatory power at the classical level. With this in mind, local states might be allowed to be as large as the most elementary entities to which a predominantly classical explanation applies. This is most probably at about the level of single molecules and is certainly not so large as a synapse or a cell. In the case of a synapse or a cell, it is possible to dissect the whole in terms of parts equally susceptible to classical analysis. Moreover, and with an eye to the subsequent critique, signalling between such parts is subject to a classical locality constraint — it requires a finite, though small, interval to transfer information from one part to another.

Very few are likely, then, to take seriously the notion that consciousness could be instantiated locally, where this is understood to mean local in the sense of *physics*, as outlined above, rather than local in the sense of *neurophysiology*. No doubt, even a *neurophysiologically local* account is to be considered with scepticism; a *physically local* explanation all the more so. Indeed, local instantiation is considered here only for the sake of completeness, and it will be more or less dispensed with after a few perfunctory comments. In the discussion of emergent accounts to follow, it will nevertheless be necessary on occasion to refer back to the concept of local instantiation and to make some further remarks regarding its credibility.

It is of course possible to consider local forms of emergence — those defined only at a single physical 'address' rather than over an extended region. It is unclear, however, what use these might be to a theory of consciousness. Local versions are not likely to provide a substrate of sufficient complexity to support an account of consciousness consistent with the observed sophistication of cognition. The fundamental unit in any classical ontology will be restricted to a finite, presumably small, number of degrees of freedom. The storage capacity of each degree of freedom will be restricted by its ability to preserve states within a given range against disruptive environmental influences like thermal noise.[6] The prospect of a local origin of consciousness would in any case appear neurophysiologically problematic, particularly in its relation to memory (see, for instance, Pribram, 1991). Unless consciousness is to be radically decoupled from the cognitive machinery — a possibility that functionalist, computationalist and complex systems theories have specifically sought to avoid — a local basis for consciousness does not appear at all promising.[7]

Nevertheless, some accounts superficially avoid the use of global states. Chalmers (1996), for instance, posits a pan-psychist interpretation of computationalism. Every local point with a computational interpretation is given a proto-conscious capacity. In this framework, however, the task of explaining the familiar brand of consciousness further requires that many disparate proto-consciousnesses be united into a single entity (James, 1890/1918) consistent with our own cognitive capacities. It is in this act of composition that global states are presumably invoked. Emergence will likely be cast in a crucial explanatory role if the account then seeks to maintain a physicalist stance with respect to this composition while avoiding brute reduction.

Extrinsic and Intrinsic Modes

If a global basis for the emergence of conscious states is for the moment granted, then any classical account can be further characterized in one of two mutually exclusive modes of description, either *extrinsic* or *intrinsic*.

An *extrinsic* mode is one that requires the postulation of an outside point of view *relative* to which the state is to be characterized. Careful accounting in an extrinsically framed theory will always expose a real or imagined observer. Any particular system might be given a multitude of extrinsic descriptions as one varies the point of view. Though different, these need not be incompatible. The 'correctness' of an extrinsic mode description is a matter to be decided independently from each point of view rather than by comparison between points of view. Assuming that each is correct within the context of its own point of view, no account is 'more correct'.

By contrast, an *intrinsic* mode invokes no point of view. States framed intrinsically are thus fully determined without any outside reference. No feature of an

---

[6] Marshall (1989) also discusses the limitations of local realizations of consciousness. His discussion is framed in terms of thermal noise blurring the distinction of adjacent states, thereby limiting the number of distinguishable states. This might be read as implying that the limiting factor is the ability of an outside observer to *discern* states (and therefore, in the vocabulary to be elaborated in the upcoming section, would be expressed in an *extrinsic* mode). But the inability of local instantiations to *preserve* particular states to an accuracy greater than that allowed by environmental factors, such as thermal noise, implies as well a limitation on complexity inherent to the local realization itself (so that limits on complexity can be read in an *intrinsic* mode as well).

[7] Note also that a local basis leaves consciousness extremely fragile. One might, for instance, expect that on such an account very minute lesions could, at least occasionally, excise consciousness entirely without incurring significant cognitive impairment. Whether this circumstance could be empirically detected might depend on the nature of phenomenal judgments in the resulting 'zombie' (Chalmers, 1996, Ch. 5).

intrinsic mode account can be made relative to the particularities of an observer's point of view.

As it is central to the distinction of extrinsic and intrinsic, the notion of a *point of view* deserves some elaboration. It is used here in a very generalized sense that should not be equated with a perspective from a given point in space, though this is naturally subsumed. The scope of its meaning can be clarified within the context of the extrinsic mode it characterizes.

In an extrinsic picture, relations between spatially separated points are made simultaneously *explicit* throughout the scope of an observer's point of view. This is what allows, for instance, the elaboration of *interpretations* defined over a *global* basis. Descriptions of function, for example, invariably involve explicit relationships between distinct points, and depend on an assessment of global context provided by an outside observer to give them meaning. Such interpretations must, hence, generally be read as extrinsic. Likewise, computational interpretations do not themselves reside in the ontology. Rather, they gain their meaning with respect to a point of view that *assigns* elements of the physical ontology to a particular computational basis. Point of view thus encompasses not only the observer's perspective (which itself subsumes numerous aspects, such as scope and special relativistic frame), but also both functional and computational interpretations.

The intrinsic mode must, on the contrary, be framed in a manner independent of outside observation. This means that it cannot draw on extrinsic elements to give an interpretation or to define a global scope within which relations become explicit. In order to qualitatively reproduce the results of our most sophisticated classical theories, we must further incorporate a locality constraint. Such a constraint, superimposed on an intrinsic picture, limits the scope of relational context (and causal contact) at each point in space to its immediate neighbourhood. There can be no direct and instantaneous communication between distinct points. All information concerning the relations of separated points is merely *implicit* in the physical structure of the aggregate system. To effect *explicit* comparisons, information must be passed, via signals of some sort, over the distance between points. Any such signals are moreover strictly limited in their speed of propagation. The maximum allowable speed is of course that of light in vacuum but, in what follows, it will be of no consequence what particular value the maximum transmission speed takes. Where the arguments invoke a locality principle, they will not derive their force from any particulars of the constraint but rather from its existence.

Relativities and the Extrinsic Mode

The invocation of an outside observer and the accompanying itinerant point of view introduces into an extrinsic picture a relativity of accounts. In an extrinsic account of conscious mental states, this relativity will have to be made consistent with the ontological character of consciousness. How this might be accomplished is best illustrated by an example. We assume in the following that the picture of consciousness that arises should be compatible with microphysical determinism and the constraints of classical science, and should take a global basis in the physical ontology. These assumptions would not appear to obviously proceed beyond those implicit in standard emergent accounts along functionalist, computationalist or complex systems lines.

1. Relativity with respect to frame

The relativity inherent in extrinsic accounts is perhaps best exemplified in the classic and instructive case of Einstein's special relativity. Consider the brain as an extended physical system. Observer A, in the rest frame of a particular conscious brain, determines the system comprised by the brain as a collection of certain events[8] with the same temporal coordinate. At each time step the emergent property identified with consciousness can be read from the characterization, and a time series of conscious mental states can be built up — call it the 'A' Series.

In a second frame, moving with some fixed velocity with respect to the rest frame of the brain, observer B similarly composes a time series of emergent characterizations of consciousness in the given brain — the 'B' Series. Each state is once again determined as a collection of simultaneous events. The notion of simultaneity, however, differs between the two frames so that none of the collections of events occurring in Series 'A' will correspond to any of those occurring in Series 'B'. It is therefore *possible*, prior to the imposition of any constraints, that the two series themselves do not correspond. It is even possible that one series contains conscious mental states while the other contains only non-conscious states.

Emergent characterizations are thus made relative to a special-relativistic frame. This might not be especially disturbing if we had no reason to require that the accounts in different frames should correspond. The relativistic mass of an extended object, for example, a feature that is determined as an emergent property of an aggregate of local particles, will vary depending on the frame and this implies no contradiction. But neither is the relativistic mass allowed an ontological status. With the advent of special relativity, the realization that the instantaneous measured mass of an object varied across frames caused the property of relativistic mass to be ousted from a fundamental position in the physical ontology, to be replaced with a new feature, invariant mass, that *was* required to be the same in all frames. Since consciousness is similarly an ontological property, it cannot be allowed that each account is equally veridical within its own frame of reference. Presumably only one series of mental states actually occurs consciously. If the 'B' Series, for instance, yields a different account of mental states from those actually experienced by the given brain, it would not seem to be possible to adopt a stance in which the 'B' Series correctly describes the mental states of that brain 'from a different perspective'. There is an ontological fact of the matter to be satisfied.

One strategy that might resolve the relativity in the two accounts is to *require* that all frames produce the same account of mental states. A solution of this kind

---

[8] The term 'event' is here used in the sense of a point in the space–time manifold designated by four co-ordinates determined with respect to a given basis.

is possible if the microphysical dynamics from which the relevant emergent property is read exhibit an *invariance* across frames. An invariance ensures that the dynamical descriptions of the emergent property remain the same no matter what relativistic transformation ('boost') of the space–time coordinates might be performed (i.e. regardless of the relativistic frame in which the states are characterized). This puts some non-trivial constraints on the sort of emergent properties that might be identified with consciousness, but it does not in principle preclude the possibility of an emergent formulation of conscious mental states within classical science. While this strategy seems entirely plausible in the context of special relativity, it will however prove much less so in the context of another kind of relativity to be discussed shortly.

There is also a second alternative available to make sense of differing accounts across frames. Here the differences are not resolved but are rather repudiated. On this approach, the only account that *must* agree with the conscious experience is the one formulated in the rest frame of the brain. No other accounts need be correct. Such a position might plausibly be defended only if an extrinsic definition of the emergent state is abandoned. Because an extrinsic formulation presumes an arbitrary external observer, it does not provide a means to select a particular point of view as 'correct'. If consciousness is *defined* on an extrinsic account, then it cannot grant priority to the case of one or another particular account from amongst all those possible in different frames. In an extrinsic theory, whether or not an emergent property is observed in a given frame is a fact that can only be 'correct' within its own frame. There is no higher criterion of truth for an extrinsic account. What is effected by this second strategy is then a shift to an intrinsic formulation in which a particular frame — in this case the rest frame — is singled out by the system itself. The emergent property is made explicit at the system level rather than at the level of an outside observer. Under such a scheme, the rest frame indeed claims priority, since the system is always trivially in its own rest frame. This line of thought, and the manner in which an intrinsic definition of consciousness might be accomplished, will be pursued in detail in a subsequent section on 'Locality and the Intrinsic Mode'.

In summary, the ontological nature of consciousness argues against ignoring the potential for discrepancies in the emergent phenomena picked out in different frames.[9] Two solutions are available: (1) resolve the multiplicity of accounts extrinsically by adopting an invariance criterion, or (2) move to an intrinsic account that rejects the multiplicity by establishing an absolute frame.

2. Relativity with respect to boundary

We now turn to a second kind of relativity evident in emergent formulations of conscious mental states. This will allow us to examine how the strategies identified with respect to relativity of frame fare in another context. Besides specifying a special relativistic frame, the external observer of extrinsic formulation also

---

[9] Note that there is reason to be concerned about such discrepancies only where emergence is specifically intended to identify an ontological feature.

delineates a *boundary* separating the 'system' — what is to be characterized — from its 'environment'. This is, of course, an arbitrary choice. However, the emergence and characterization of properties that explicitly recognize information or relationships available only at the global level will be sensitive to this specification. In order to recognize properties emerging from a global state, the delineation of the system must meet the specifications carried by the operative definition of 'global'. We would not, for instance, expect to find the emergent property of 'automobile-ness' within a carburettor. The system (in this case, the carburettor) has not been defined in accordance with the requirements of a global state (a car) from which 'automobile-ness' might emerge. It is thus only in the context of a specific choice of the system boundary that the character of an emergent property is determined. This is not to say, at least in the usual case, that the emergent property prescribes a particular system. Rather it says that whether or not the property is observed, and its particulars, will depend in part on the boundary chosen to circumscribe the system.[10]

The manner in which a relativity with respect to the definition of the system boundary comes into the extrinsic determination of emergent properties can be made more clear with an example. A paradigm case of an emergent property often cited in the literature is 'liquidity'. 'Liquidity' is a higher-order feature of certain systems encompassing at the micro-level large numbers of water molecules. Evidently the property is compatible with microphysical determinism, but it does not reduce to a property of isolated molecules or their reducible relations. Neither is it necessary for the causal dynamics of systems of water molecules to be invested with an explicit account of 'liquidity'. When and where it is observed, it is realized as a global property of an observer-defined 'system'. It is, for instance, not observed within a glass of water if the scale of the system specified is at the level of individual molecules. On the other hand, the property *is* observed in a system encompassing all of the water in the glass. It is *also* observed in smaller systems containing only part of the water in the glass and in larger systems containing several glasses of water. No boundary is inherently preferred by the phenomenon of 'liquidity'. Hence, any discussion of the emergence of 'liquidity' must come with an identification of the system from which it is presumed to emerge. What should be noted is that there is no contradiction in allowing all of these systems, variously defined, to have the property of 'liquidity'.

The phenomenon of consciousness would seem to stand in marked contrast. Imagine several different accounts of the emergence of consciousness from the physical goings-on in a particular brain. Each one defines the boundary of the physical system under consideration somewhat differently. If we are to understand emergence in the same way that it is understood for phenomena like 'liquidity', then the emergent property should be expected to manifest in a great many of these accounts. Recall that the designated property must be one whose definition is flexible enough to be discerned in the brain of *any* conscious individual — a population that presumably shows a great deal of variation

---

[10] Except of course in the trivial case of an emergent property that is identically characterized, regardless of what system (i.e. the entire universe or any of its parts) it is taken to be a feature.

— at *every* moment of their conscious awareness. Recall also that it is here our goal to consider only extrinsic accounts, formulations in which the criteria of emergence cannot be specified *in terms of* particular boundaries (particular points of view). With these things in mind, it is difficult to imagine how the relevant emergent feature might be defined such that it would be satisfied only if the system boundary in a given brain is chosen in a very particular way. This would require that no other system, though it may differ from this first system by only the smallest or subtlest of changes in its boundary, could be allowed to manifest the emergent property. Yet the definition of this emergent property would nevertheless be satisfied by systems in every other conscious brain, though these surely bear substantial physical differences to the first.

Hence, we assume that the emergent property identified with consciousness will be manifest in numerous systems, differing in the precise determination of their boundaries. As the system boundary is varied across all of these accounts, the content of consciousness would then also seem to have to vary correspondingly if there is to be a 'cognitive coherence' (Chalmers, 1995) between the conscious state and the informational base of the defined system. But only one consciousness actually emerges from the studied brain. Only one conscious state will actually be experienced per brain per moment. Without knowing what exactly are the contents of that particular mental state, we know that those contents are fixed. There are matters of fact in the mental ontology with which no contradiction can be tolerated.[11] A single mental state cannot simultaneously satisfy a multitude of *different* ascriptions of content across system definitions.

The singularity of conscious experience means that it must be determined independently of the way an external observer extracts, from the relevant facts about the system and a specification of its boundary, an interpretation that merely *renders* consciousness from the point of view of the given boundary selection. This independence might be made consistent with an emergent account by adopting one or the other of the same two strategies that were available in the case of relativity with respect to frame. On the first strategy, one envisions that there exists an invariance property. The postulated invariance would ensure that the *same* mental state would be read from the dynamics regardless of which 'system', amongst those that satisfy the criteria for emergence, is assumed by the external observer. The problem here is that, in order to achieve the desired equivalence of emergent mental states, the invariance must radically homogenize mental content across very different ways of carving up the brain into physical systems. At the level of systems spanning large parts of the brain, there is already considerable room for a vast multiplicity in the number of ways that given functional or computational criteria can be met. Each different way represents a system with distinct informational content, yet each of these must determine identical conscious content. Moreover, many of the emergent properties currently being discussed (Skarda

---

[11] It is perhaps necessary to state that this makes no assumptions concerning the conscious agent *vis-à-vis* infallibility of introspection or omniscience concerning mental states. There is no claim here about *access* to all mental states, nor is there any *judgment* invoked with respect to the contents of consciousness. Justification comes from brute matters of fact concerning the ontological *existence* of conscious mental states.

and Freeman, 1990; Freeman and Skarda, 1991; Churchland, 1995; Greenfield, 1995; Port and van Gelder, 1995), particularly those in complex systems theories of consciousness, could manifest even within isolated modalities and functional domains. Not only would this multiply the number of systems subject to the invariance, but it would also mean that many of them would be physically disjoint.

An invariance of the postulated kind must then ensure the equivalence with respect to content of all such systems, variously defined. But specific information, that might be available within certain system definitions, would presumably not be available on all other possible definitions. It is thus difficult to see how all such accounts could lead to equivalent mental states. How will it be possible to reconcile a single experience of consciousness with a definition of emergence that sees myriad different systems within the brain all manifesting the specified property? Each is required to carry the same mental content, though each is invested with different amounts of information. It seems impossible, moreover, to disqualify instances of emergence that arise when the extrinsic scope is expanded *beyond* a single brain,[12] a point raised by Rosenberg (1997, Ch. 9). The required invariance would then seem to imply that the separate instances of the emergence of consciousness in two distinct brains must each yield equivalent content to that of the single instance that emerges from the system composed of both brains. Whether or not the information available in one of these brains is also available to the other, the invariance demands that the emergent mental states should be equivalent. Such an invariance leads to the absurd conclusion that it should be impossible to discriminate the consciousnesses of different individuals!

To avoid assigning mental content to physical systems where it cannot possibly be justified on informational grounds, one might stipulate that the invariant mental content should be limited to that carried by the system which *minimally* meets the criteria for consciousness. Immediately we should note that this system is not guaranteed to be unique. A 'minimally conscious' system, one for which any reduction in physical scope whatsoever would extinguish the possibility of a conscious state being manifest, might be determined in a vast number of ways. The uniqueness of the minimal system could only be assured if its spatial extent were entirely contained in every other system definition in which the property is manifest. Only if it is unique could the multiplicity of different extrinsic formulations of the conscious state be given *any* interpretation of content consistent with the informational scope of each.

That it might be implausible that such 'minimally conscious' systems should invariably be embedded as invariant subsets in larger systems, also conscious, is suggested if one imagines pruning one of these larger system definitions by one neuron.[13] There will in general be a vast number of ways to do this, each resulting

---

[12] Since the issue is itself one of scope, the assumption of a particular extent to the 'brain' cannot be made prior to deciding the issue without incurring circularity.

[13] This presumes that the neural level is the appropriate one at which to frame classical models of consciousness. The discussion of pruning would in fact be better phrased at the level of individual molecules (or whatever might constitute the base in the relevant classical physical ontology), where such assumptions could be avoided, but is left in terms of neurons to make contact with standard theories of consciousness.

in a new system.[14] Some of the new systems may fail to meet the criteria of emergence. Nevertheless, if it is going to be possible to credit many and considerably variable brains with giving rise to conscious states over a wide range of times, one should expect that at least some, possibly many, might preserve the emergent character of the original. While each of the pruned systems is wholly contained in the original system, none are wholly contained in any of the other pruned systems. Iterations of the pruning process will most likely lead, in the absence of further restrictions, to a multitude of overlapping, perhaps even disjoint, systems at the 'minimally conscious' level. A unique 'minimally conscious' system might be recovered *only* where that system is wholly contained within the overlap of *all* larger systems, assuming that pruning identifies *no* other minimal systems.

The additional stipulation that the minimal system should always be found as a unique and invariant subset of all larger systems deemed to exhibit consciousness might resolve any potential problems. While it is perhaps counter-intuitive that the mental content of *any* conscious state should be invariably equivalent to the minimum content necessary to maintain consciousness, this observation does not undermine the coherence of a strategy formulating content in terms of minimal systems. What makes the pursuit of such a course untenable is rather that it constitutes an *intrinsic* solution to an *extrinsic* problem. According to the line of reasoning that has been developed, it is imagined that the phenomenon itself, in the manner of *intrinsic* formulations, singles out a unique and invariant 'minimally conscious' system. The determination of mental content can no longer be made purely from the consideration of an arbitrarily defined boundary and the system contained therein (as well as the relevant boundary conditions), as required in an extrinsic formulation. Instead, a particular system that minimally meets the criteria for the emergence of consciousness must be selected from the ensemble of *all* such systems. In order for content to be consistently assigned in each of these systems while yet giving rise to only one conscious state, there must exist a mechanism by which to inform all larger systems of the identity and content of the minimal system. Whether or not such a mechanism can be found is not at issue here. What is important is that a strategy which singles out the relevance of particular systems in this manner is not available in an extrinsic picture. Conversely, if we are willing to consider intrinsic formulations, there is no need to introduce the invariance principle that spawned the entire problem.

---

[14] If the original system is itself 'minimally conscious', then it will of course not be possible to conduct any pruning. Since we are interested in determining how the 'minimally conscious' systems are embedded in larger systems, this should only be the case if *no larger systems exist that are also conscious*. In this event, it is to be imagined possible to define criteria for emergence, *on an extrinsic account* that does not allow the phenomenon itself to single out particular boundaries, that would fail to be met if even a single neuron were added or removed from the brain. Moreover, since the pruning might just as well be conducted at the molecular level, the position must be read as asserting that not even a single molecule could be added or removed without losing the conditions for emergent consciousness. It is unclear then how anyone might remain conscious for more than a brief interval under such a rigid extrinsic definition of emergence, since molecules are regularly delivered to the brain across the blood-brain barrier and removed from the brain by the cerebrospinal fluid. It is even more unclear how this position could be maintained while simultaneously avowing that the given criteria could nevertheless be met in the brains of other conscious individuals.

If the invariance principle is adhered to on a rigorously extrinsic programme, then it must be unrestricted by stipulations that refer to intrinsically determined concepts like 'minimally conscious' states. This then opens the floodgates so that, in considering any given extrinsic system, almost any content might be countenanced, and the account therefore would appear to be under-specified. The only remaining avenue to pursue is one according to which extrinsic criteria so stringently specify the conditions of emergence that consciousness should be seen to arise only in one particular system. But while it is possible in practice that this might happen to be always the case, it is impossible to impose it in principle. To do so would be to *guarantee* that there should never be more than one system on the basis of which mental content should be determined, that a discrepancy should never occur between the information contained in several candidate systems; and this must hold at every conscious moment for every individual. The extrinsic approach is incapable of making such promises. In an extrinsic picture, whether consciousness emerges is a matter determined relative to each point of view and the system it defines. It cannot be decided by a contest of different points of view.

Since an extrinsic approach to consciousness does not appear to be tenable, we might explore an alternative strategy. The required independence of consciousness from the particularities of an external description might be granted, without invoking a disquieting invariance, if the system internally recognized that the conditions for emergence were met only with regard to a very specific definition of the system and no other. If consciousness is to be explained as an emergent property of a physical system, it appears then that the 'system' must have an intrinsic capacity to delineate itself. There is, on this view, a 'preferred' boundary. There seems to be no corresponding sense in which it is required of other emergent phenomena in the classical domain to exhibit such a capacity to inherently 'know' the exact scope of the system necessary to satisfy their conditions of existence. The requirements of an intrinsic model of emergent consciousness are thus expected to be unique. We turn now to an examination of such intrinsic formulations of conscious mental states.

Locality and the Intrinsic Mode

The natural reading of most emergent theories is an extrinsic one. Functional attributions, computational interpretations and in general any descriptions invoking relationships that are not explicit in the physical ontology, would seem to require a point of view to fix their scope and sense. Nevertheless, emergentism might be enlarged to make room for the possibility of an intrinsic reading. Some in the functionalist camp, for instance, have challenged the interpretation of function as an extrinsic feature. Chalmers (1995) characterizes the broadest view of function that has been employed by functionalists as 'any causal role in the production of behaviour that a system might perform'. On this definition, functionalism does not appear inherently incompatible with an intrinsic definition of consciousness. Similarly expanded definitions might render computationalism and complex systems theories amenable to intrinsic interpretation as well.

Despite overcoming the inherent flaws of the extrinsic account, intrinsic accounts of emergentism still seem unsalvageable within the context of classical science alone. This is because the states from which emergent properties are read are determined as physically extended entities (i.e. the emergent properties take a global basis). If emergent properties are to have causal consequences, the global state of an entire system must continuously inform the local dynamics, where all the causal machinery of the classical worldview is located. If the local level is not seen to be the sole repository of causal dynamics, any account *compatible* with naturalism and the principle of microphysical determinism must still allow that all causal processes have an interpretation at this level. Thus, the global state must update local dynamics as to the current whereabouts of the boundary, in order to establish the physical extent of the emergent unit that it constitutes. Even if, on an epiphenomenal view, emergent properties are not regarded as causally efficacious, some account of the relevant boundary must be explicitly available to facilitate the realization of the corresponding conscious state, rather than one associated with a different boundary.

This, however, would appear to be impossible. Classically, a universal locality principle forbids that an explicit representation of the boundary can be made available throughout an extended global system simultaneously. The principle holds that communicating explicit information over physical distances, *however small*, does not occur instantaneously. The enforcement of a locality principle prevents classical science from recognizing intrinsically defined systems. This is not, it should be noted, merely a problem in practice, related to the particular speed of signals or of computation, but an issue in principle, related to the notion of simultaneity in the classical worldview (which includes relativity, but not quantum theory).

We can play out the consequences of a locality constraint in the context of a 'toy' model comprising a group of neurons seeking to form an intrinsically-defined global state. If the neurons are to indicate their membership in this global state in a manner concordant with the precepts of classical science and requiring no extrinsic interpretation, then some signal must be transferred between them to indicate which neurons will instantaneously enjoy membership. Thus neuron 'A' transmits a signal to other neurons to confirm its membership in the global state as of this moment, call it t_1. Regardless of the form it takes, this signal is transmitted subject to a locality constraint, restricting the speed of its propagation to some finite value. It, therefore, cannot be received by any other neuron until some later time, say t_2. It is at this time that a second neuron, neuron 'B', has assurance that at time t_1 neuron 'A' was to have been considered part of a global state. Of course, the signal has arrived too late to have any causal consequences concerning the establishment of a global state at time t_1. We might nevertheless imagine that a global state could be established at time t_2 on the basis of information contained in this signal, supplemented by the additional assumption that neuron 'A' is *still* to be considered part of the global state. The problem here is that the signal in fact indicated nothing about the status of neuron 'A' at the later time t_2. In fact, neuron 'A' *may* in the interim have ended its participation. To circumvent this difficulty we might go on to imagine that neuron 'A' included in the signal sent at time t_1 the information that it should *still* be considered for membership in any global state that might be established at time t_2. This however would require neuron 'A' to anticipate, by a sort of microscopic prescience, any causal influences acting upon it in the interim. This sort of scenario must presumably violate any kind of causality that might be operational in a classical framework. Perhaps then neuron 'B' might be 'justified' in the *assumption* that neuron 'A' continues in its quest to become involved in a global state, if neuron 'A' has persisted in sending such signals over a long period of time. This will also not do to establish a global state, however. It constitutes an extrapolation by neuron 'B' of the data received from neuron 'A'. The basis of this extrapolation must be information stored by neuron 'B' so that it establishes only a *local* state at neuron 'B'[15] that does not involve neuron 'A'. We might imagine, for instance, that neuron 'A' was obliterated immediately after the transmission of its final signal, prior to the extrapolation by neuron 'B'. Less drastic occurrences would, of course, produce the same result.

In the last of the scenarios above, information about global aspects was allowed to accrue at a local site. In this way global properties could become causally empowered with respect to microphysical determinism. This escape is theoretically constrained, however, by the ability of classical systems to store information locally. More importantly, there appears to be little evidence that a 'central accumulator' of the hypothesized sort actually exists in the brain. The neurophysiological problem has always been that the requisite information for consciousness is nowhere to be found together in one place. If this is neurophysiologically dubious, it is all the more so in the vastly more restrictive sense of physics, where the relevant scale is about the size of a single molecule.

Since it is not convenient in the context of establishing intrinsic global states, to strictly distinguish the two times t_1 and t_2, one might imagine introducing something like a 'fudge factor'. This would allow one to ignore the sequencing of causal prescriptions for brief intervals, to treat the two distinct times as if they were effectively one, should they occur close enough together. Such a strategy might be adopted on the view that, for distances measured within the brain, the length of time between t_1 and t_2 will be negligible. Though a 'fudge factor' is obviously not part of standard classical explanation, one might nevertheless try to implement it in a principled and consistent manner. For instance, it might be made a universally applicable element of the classical canon. Alternatively, it might be admitted singularly as an accompaniment of consciousness.

On the first hypothesis, the 'fudge factor' is built into the objective notion of time, so that states might be allowed to enjoy global bases extending over space-like intervals. It would then seem impossible in principle to accurately measure

---

[15] Again, the discussion is framed at the level of neurons to make contact with standard theories, and should in fact be framed at the most fundamental level in the classical ontology. Thus the 'local state' to which the statement refers must actually be even further localized within the structure of the neuron, exacerbating the difficulties.

times smaller than those necessary to traverse the relevant neurophysiological structures. Even if we opt for a smaller scale than would tend to be supported neurophysiologically — say, one millimetre — the traversal time at the speed of light is already as long as three picoseconds. Time is the most accurately measured of physical quantities, and is routinely measured to this accuracy in atomic clocks.

On the second hypothesis, the 'fudge factor' is not taken to be constituent of fundamental laws, but rather is a contingent fact in the brain. The question then is how the 'fudge factor' establishes itself *prior* to the determination of the conscious state in terms of which it derives its meaning. In particular, there must be some account, preferably not *ad hoc*, of where and when the 'fudge factor' can come into play, and the extent to which it blurs the distinction between otherwise distinct times. This account must not, to avoid circularity, depend on the determination of intrinsic global states. It likewise cannot rely on extrinsically determined features or properties if it is to preserve the intrinsic character of the explanation; and there does not appear to be a further option to pursue.

If it is not possible to accumulate something with the full complexity of consciousness, then perhaps less extensive global states might yet be constructed in a locally intrinsic manner. These might then become appended in some invariant way to form a single conscious state. This strategy involves a recurrence, now at one remove, of the problems inherent to extrinsic formulation (of ontological subject matter), problems once again related to the determination of an appropriate boundary, here determining the level at which appending must halt.

More importantly, however, the suggestion tacitly assumes the existence of a 'place' in which the process of appending can proceed. Since, by hypothesis, the reduced global states are already endowed with an emergent property identified with mental features, we might term this theatre a 'mental space'. This mental space is what must establish and maintain ordered relations between all the various mental 'bits' as they are composed. What is problematic about this conception is that a mental space cannot be merely an abstract, theoretical or mathematical construct. It is as much an ontological forum as is physical space. Thus, we are not free to assume that it comes with a pre-existing structure, as we do when we employ an imagined space in the way of a device or tool. While the structure of a mathematical space begs no explanation, there will be matters of fact about the structure of an ontological space — whether it be physical space or the posited mental space — that do require explanation. But since the mental space *constitutes* the (mental) locality relations by which mental features are to be ordered, these locality relations are not then available to an explanation of the structure of mental space itself.

The inherent difficulties can be illustrated by considering the analogous situation, encountered in cosmology, of explaining how *physical* space–time achieves its structure without implicitly assuming that this occurs within another, logically prior, space. The naïve approach simply arranges in some kind of succession whatever primitive elements might compose physical space–time. This process unwittingly invokes a further, underlying space, necessary to give sense to the process of arranging but whose structure is then left unexplained. For this reason, cosmological theories of space–time structure are necessarily non-local. A local account would not even make sense, since locality relations *exist* only once the space has been structured — they are *to be explained*, not used in the explanation.

A mental space must be similarly constructed[16] in a non-local manner (but note that 'non-local' here refers to non-local in *mental* space). Thus all the relations necessary for its construction must be provided altogether and at once. No piecemeal process of composition will be possible. The situation bears a metaphorical resemblance to assembling a jigsaw puzzle. It is usual to put together such a puzzle piece by piece, allowing some flat surface to preserve the relations (provide the locality relations) of that portion already assembled while additional pieces are attached. Without the benefit of a surface on which to work (without a pre-existing structured space), assembling the puzzle becomes much more difficult, since all the relational information must be provided explicitly and at once, say by holding all the pieces together in one's hands.

To construct the postulated mental space, the classical paradigm provides us with two options. Either the requisite relations could be encoded locally, or in a distributed fashion (in both cases here, we refer to *physical* space). In the former case, the information might indeed be made available for an 'altogether and at once' act of composing. But the amount of information required to specify explicit relations between each element of the mental space and every other element, even if this is not required at the most primitive level, rapidly outstrips the capacity for local storage in physical space at biological temperatures. Allowing for only the pair-wise relations amongst a set of N 'items' in consciousness, and restricting these to be each specified by only six bits, the number of states required for the encoding is $2^{3N(N-1)} \approx 10^{18}$ for $N=5$. To compare, specifying the items themselves with six bits each, and ignoring their relations, would require only $2^{6N} \approx 10^9$ states for $N=5$. The discrepancy between these two figures is exponentially exacerbated with each increment in N. Noting further that relations need not only be spatial but might also be expressed along visual dimensions like colour, as well as in other modalities and in a conceptual (as opposed to perceptual) domain, it seems clear that a six bit encoding almost certainly underestimates the requirements. No local representation (as, for example, the states of a single molecule) can accommodate a state space of the requisite size while contending with the disruptive effects of thermal noise at room temperature (Marshall, 1989). We are thus led to investigate distributed representations.

In the distributed case, the locality principle forbids simultaneous access to distributed information, however small are the distances involved. The relations *implicit* in a distributed account are never allowed to become *explicit*. The global (distributed) level at which outside observers (extrinsically) access explicit relations cannot simultaneously inform the local dynamics throughout a spatially

---

[16] However a mental space, at least in a naturalistic theory, need not be constructed *ex nihilo*, and on this point its conditions of explanation diverge from those of physical space in cosmology. For example, no account of the *existence* of information itself is needed in a theory of consciousness. The relational information necessary to order mental space is provided, in a physicalist account, by the physical ontology.

extended region. The level at which relations become explicit in an extrinsic account is therefore without causal consequence. What information can be made available simultaneously at a local level in an intrinsic picture will be far too scanty to form the foundation on which the hypothetical space might be constructed.

It is worthwhile noting, in this context, that the distributed processing at the basis of many neural network-based theories of consciousness (see, for instance, Edelman, 1989; Port and van Gelder, 1995) does not constitute the kind of non-local connection that would be required to circumvent the locality principle in an intrinsic theory of consciousness. What is accomplished by distributed processing is the *disposition* of downstream neural dynamics. It is only in these dispositions (and the resulting behaviour) that the implicit order of distributed processing is made explicit. These dispositions, however, give us no indication why the physical state should be accompanied by a distinct global state. The 'attractors' of complex systems theories and the synchronous firing in temporal binding theses, for instance, are features of a system of variables extrinsically determined and truncated at the discretion of an observer. We have already seen that it is not possible to give a consistent account of conscious states according to extrinsic criteria. Global properties like these, that remain distributed throughout an extended region of space, cannot become unambiguously explicit without the aid of an observer unless the locality principle is contravened. In classical accounts, the means to establish intrinsic criteria for emergent properties like 'attractors' and 'neuronal gestalts' — criteria that do not depend on an arbitrary outside assignment — must always involve signalling, and signalling is subject to the locality principle. Hence distributed properties would seem to always reduce to the sum of local physical properties, leaving no basis on which to render conscious states that are ontologically distinct from their purely physical counterparts.

One might then imagine that it is simply not necessary to give an explicit account of the global properties of an intrinsic system, and in particular of its boundary. Our discussion in this article has been premised on the notion that conscious states must be given an *explicit* physical basis. This assumption is, for example, contradicted by identity theories that would allow an *implicit* identity between conscious states and their physical counterparts. On such an account, however, it is not determined within the physical basis that any conditions of satisfaction for conscious states have been met. Rather, the necessary identities, between each instance of a conscious state and an associated physical state, are built in at the level of physical law. This solution removes the problem of consciousness from the domain of science. Science is relegated to a taxonomic role, tabulating the various physical states with which conscious states are putatively associated. There would appear to be no opportunity in this enterprise for the kind of generalizations from which science draws its explanatory power, since all of the correlations between physical and phenomenal must reside at the most fundamental level. If they do not, then one can ask how it might be determined in the physical substrate that particular physical states are sufficient to give rise to conscious states; and that would require an explicit treatment of the physical realization of consciousness.

Conclusion

To be useful in bridging the gap between the mental and physical ontologies while yet conforming to the standard principles of classical science, the notion of emergence must be adopted in a form that incorporates ontological novelty and accommodates microphysical determinism. Where emergent accounts of consciousness are further characterized as extrinsic, they must be reconciled with a variety of relativities with respect to external observers. Relativity with respect to frame can be redressed by requiring that the relevant emergent property be frame invariant. No such strategy is, however, available in the case of relativity with respect to boundary, at least not without absurd consequences. If consciousness then requires an intrinsic definition, the local dynamics must be continuously informed of the global positioning of the boundary between system and environment, if the account is to be causally empowered. Even if consciousness is not causally efficacious, the determination of a global state still requires that the specification of the relevant boundary be made in some way explicit. In either case, this is incompatible with locality constraints that apply throughout classical science.

Instead, entirely local instantiations could conceivably be instantiated where a vast number of local degrees of freedom are sufficiently shielded from thermal noise. There seems little reason to pursue this possibility, at least where the full complexity of consciousness is involved. One of the most persistent conclusions of neurophysiological investigation has been that there is no confluence of informational content. A *physically* local realization is yet more unlikely than a neurophysiologically local one (see p. 103 above).

Consciousness might yet be realized, not within the context of a single global state, but through the conjoining of a plurality of mental features, each sufficiently simple to allow local instantiation. This procedure tacitly invokes a structured mental forum in which ordered relations between features can be established and maintained. Such a forum must be given ontological account. Since there are no pre-existing locality relations by which to compose mental content, the construction of a 'mental space' cannot occur piecemeal. This imposes the need for explicit relations between all the compositional elements, and vastly increases the amount of information necessary to build mental states. Once again, the requirements are not realistically compatible with the constraints imposed by locality in physical space.

All of this leads to a theory of consciousness in one of several different directions. Clearly one can deny consciousness the ontological status that it has been accorded here, the consequence being a behaviourist or radically eliminativist theory. Denying the existence of consciousness should not, however, have strong appeal for those with immediate evidence to the contrary. Alternatively, abandoning the idea of a naturalistic theory will extract one from the conclusion. The available options along this route have not been considered generally palatable: either one abandons parsimony and develops separate 'rules of conduct' for mental and physical realms and again for their interaction, or one is driven to some version of epiphenomenalism.

One might also abandon the search for explicit physicalist realizations and adopt the implicit account of an entirely reductive physicalism. The difficulty here is that the solution allows us no deeper understanding of the workings of our universe. There are no criteria by which to assess in the physical substrate whether conditions of satisfaction have been met for the determination of a conscious state. Rather than obtaining explanation, this approach eschews the problem by off-loading it, in its entirety, to fundamental law. No facts remain of which the theory might be predictive. This sort of strategy is available to the solution of almost any kind of problem, but it is not generally constitutive of a scientific approach. Science normally demands substantial return for each investment at the level of fundamental law. It is not clear, in this case, that any return at all can be expected from the considerable deposit required by standard identity theories.

It is also possible that emergence is not the concept that will render an account of conscious ontology. Perhaps whatever notion of emergence is useful to an explanation of consciousness cannot be made compatible with a principle of causation expressed in terms of microphysical determinism. In the first case, a characterization of the principle that will provide the bridge from physical states to conscious ontologies must preclude any further progress on the problem from a theoretical standpoint. In the second case, the manner in which non-standard conceptions of causation might alleviate the concerns addressed in this paper should be made clear in the context of particular theories of consciousness, in order to assess the damage incurred to the scientific edifice by the shift to a novel form of causation. Theories in terms of 'downward causation' are not, at present, sufficiently elaborated to be able to determine the conditions of satisfaction that must hold if formal-final causes are to be allowed into the causal chain. It is also not clear that, once spelled out, these theories will be sufficiently differentiated from the emergentism discussed in this article to evade the arguments presented here. If they are, they must face the additional hurdle of demonstrating that an alternative account, which allows for 'downward causation', can be given of the body of accumulated scientific results. Process philosophy, as well, requires substantial elaboration before it is susceptible to analysis, both in terms of answering the issues raised in this article and in providing alternate account of 'state' science.

Finally, the constraints imposed by locality might be evaded in theories involving a non-trivial component lying outside the bounds of classical science. The non-computational processes considered by Penrose (1994), for instance, might qualify in this respect, and the remarks given here are consonant with conclusions he draws from his Gödelian argument. A solution that allows the introduction of quantum coherent states might make considerable progress in answering our objections, particularly if, as seems likely, it turns out that the paradigm of extrinsically defined states could be evaded in this context. Obviously this route poses considerable problems in practice, as it is far from apparent what sort of role, if any, such states could play in the messy environment of the brain or how they might be related to neural explanations of cognition (for some possibilities in this direction see Hameroff and Penrose, 1996; Jibu *et al.*, 1994; Pribram, 1991). But while it is clear that conscious states are intimately related to the functioning of neural systems, this observation does not preclude the possibility that other processes occurring in the brain might hold an essential piece of the puzzle.

Acknowledgements

This work was supported in part by a Science and Technology Agency (STA) fellowship. We would like to thank D. Chalmers, D. Douglas-Brown, G. Globus, K. Pribram, W. Seager and Y. Stapp for helpful discussions and insightful commentary.

References

Ali, S.M., Zimmer, R.M. and Elstob, C.M. (1998), 'The question concerning emergence: implications for artificiality', in *Computing Anticipatory Systems: CASYS — First International Conference*, ed. D.M. Dubois (AIP Conference Proceedings) **437**, pp. 138–56.

Bohm, D. and Hiley, B.J. (1993), *The Undivided Universe: An Ontological Interpretation of Quantum Theory* (London: Routledge).

Campbell, D.T. (1974), ' "Downward causation" in hierarchically organised biological systems', in *Studies in the Philosophy of Biology*, ed. F.J. Ayala and T. Dobzhansky (New York: Macmillan).

Chalmers, D.J. (1995), 'Facing up to the problem of consciousness', *Journal of Consciousness Studies*, **2** (3), pp. 200–19.

Chalmers, D.J. (1996), *The Conscious Mind: In Search of a Fundamental Theory* (Oxford: Oxford University Press).

Churchland, P.M. (1995), *The Engine of Reason, the Seat of the Soul: A Philosophical Journey into the Brain* (Cambridge, MA: MIT Press).

Churchland, P.S. (1995), 'Take it apart and see how it runs', in *Speaking Minds: Interviews with Twenty Eminent Cognitive Scientists*, ed. P. Baumgartner and S. Payr (Princeton, NJ: Princeton University Press).

Collier, J.D. (1998), 'The dynamical basis of emergence in natural hierarchies', in *Emergence, Complexity, Hierarchy and Organization: Selected and Edited Papers from the ECHO III Conference, Acta Polytechnica Scandinavica*, ed. G. Farre and T. Oksala (Espoo: Finnish Academy of Technology).

Crick, F. and Koch, C. (1990), 'Towards a neurobiological theory of consciousness', *Seminar in the Neurosciences*, **2**, pp. 263–75.

Dennett, D.C. (1991), *Consciousness Explained* (Boston, MA: Little, Brown and Co.).

Edelman, G.M. (1989), *The Remembered Present: A Biological Theory of Consciousness* (New York: Basic Books).

Emmeche, C., Koppe, S. and Stjernfelt, F. (1997), 'Explaining emergence', *Journal for General Philosophy of Science*, **28**, pp. 83–119.

Freeman, W.J. and Skarda, C.A. (1991), 'Mind/brain science: neuroscience on philosophy of mind', in *John Searle and His Critics*, ed. E. Lepore and R. van Gulik (Oxford: Blackwell), pp. 115–27.

Greenfield, S.A. (1995), *Journey to the Centers of the Mind* (New York: W.H. Freeman).

Hameroff, S. and Penrose, R. (1996), 'Conscious events as orchestrated space–time selections', *Journal of Consciousness Studies*, **3** (1), pp. 36–53.

James, W. (1890/1918), *The Principles of Psychology*, Vol. I (New York: Henry Holt & Co.).

Jibu, M., Hagan, S., Hameroff, S.R., Pribram, K.H. and Yasue, K. (1994), 'Quantum optical coherence in cytoskeletal microtubules: implications for brain function', *BioSystems*, **32**, pp. 195–209.

Klee, R. (1984), 'Micro-determinism and concepts of emergence', *Philosophy of Science*, **51**, pp. 44–63.

Marshall, I.N. (1989), 'Consciousness and Bose-Einstein condensates', *New Ideas in Psychology*, **7** (1), pp. 73–83.

Penrose, R. (1994), *Shadows of the Mind* (Oxford: Oxford University Press).

Polanyi, M. (1967), *The Tacit Dimension* (London: Routledge and Kegan Paul).

Port, R.F. and van Gelder, T. (ed. 1995), *Mind as Motion: Explorations in the Dynamics of Cognition* (Cambridge, MA: MIT Press).
Pribram, K.H. (1991), *Brain and Perception* (New Jersey: Lawrence Erlbaum).
Rosenberg, G.H. (1997), *A Place for Consciousness: Probing the Deep Structure of the Natural World* (Bloomington, IN: Indiana University Press).
Scott, A. (1995), *Stairway to the Mind: The Controversial New Science of Consciousness* (New York: Springer-Verlag).
Skarda, C.A. and Freeman, W.J. (1990), 'Chaos and the new science of the brain', *Concepts in Neuroscience*, **1**, pp. 275–85.
Sperry, R.W. (1980), 'Mind–brain interaction: mentalism yes, dualism no', *Neuroscience*, **5**, pp. 195–206.
Sperry, R.W. (1992), 'Turnabout on consciousness: a mentalist view', *Journal of Mind and Behaviour*, **13**, pp. 259–80.
Whitehead, A.N. (1929/1985), *Process and Reality: An Essay in Cosmology* (London: The Free Press).

Todd E. Feinberg

Why the Mind is Not a Radically Emergent Feature of the Brain[1]

*In this article I will attempt to refute the claim that the mind is a radically emergent feature of the brain. First, the inter-related concepts of **emergence**, **reducibility** and **constraint** are considered, particularly as these ideas relate to hierarchical biological systems. The implications of radical emergence theories of the mind such as the one posited by Roger Sperry, are explored. I then argue that the failure of Sperry's model is based on the notion that consciousness arises as a radically emergent feature 'at the top command' of a non-nested neurological hierarchy. An alternative model, one that avoids the dualism inherent in radical emergence theories, is offered in which the brain is described as producing a **nested hierarchy of meaning and purpose** that has no 'top' or 'summit'. Finally, I will argue there remains a non-reducible aspect of consciousness that does not depend upon radical emergence theory, but rather on the mutual irreducibility of the subjective and objective points of view. This irreducible aspect of consciousness can be understood as the **non-mysterious** result of brain evolution and normal neural functioning.*

I: Introduction

'How does the brain, which is composed of billions of individual neurons, create a unified mind?' Science writer John Horgan, in his recent and provocative book *The Undiscovered Mind* called this problem the 'Humpty Dumpty Dilemma':

> This conundrum is sometimes called the binding problem. I would like to propose another term: the Humpty Dumpty dilemma. It plagues not only neuroscience but also evolutionary psychology, cognitive science, artificial intelligence — and indeed all fields that divide the mind into a collection of relatively discrete 'modules,' 'intelligences,' 'instincts,' or 'computational devices'. Like a precocious eight-year-old tinkering with a radio, mind-scientists excel at taking the brain apart, but they have no idea how to put it back together again (Horgan, 1999).

---

[1] This article is adapted with permission from the book *Altered Egos: How the Brain Creates the Self*, by Todd E Feinberg © 2001 by Oxford University Press, Inc.

Figure 1.
Many models of the mind envisage the brain as a hierarchy in the shape of a pyramid. The many parts of the brain that contribute to the mind make up the base of the pyramid. These parts are combined and organized to create higher levels of the hierarchy. Suddenly and somewhat mysteriously, a unified mind and self is supposed to 'emerge' at the top.

The problem of mental unity poses a real challenge for any neurobiological theory of the mind. If all brain regions that contribute to consciousness can be enumerated and tallied as if they were computer modules, how are they integrated so that we exist as unified, single selves? What is it about the brain that creates the subjective sense that we possess a single and unified point of view, an inner 'I'? What keeps the neurons of our brains from going off in their own directions?

One potential solution to the 'Humpty Dumpty Dilemma' is that the mind, somewhat mysteriously, *emerges* from the brain. In this line of reasoning, consciousness, as an emergent feature of the brain, extends beyond, or is 'more than the sum of the parts' of the brain. And since the mind transcends the brain, while the brain may be divisible, consciousness can nonetheless emerge *unified* from the brain-like the eye that emerges from the top of the pyramid on a dollar bill (Figure 1).

The idea that the mind emerges at the highest levels of the nervous system at the pinnacle of the neural hierarchy is appealing for other reasons as well. Consciousness seems to entail the 'highest' and most advanced forms of cognitive processing, and the phylogenetically most advanced regions of the nervous system, for example the frontal lobes, are indeed situated 'higher' on the neuroaxis when compared with regions, such as the midbrain, that subserve more automatic behaviors. It therefore seems reasonable to suppose that 'higher cortical functions' requiring the involvement of consciousness might emerge within the most advanced and evolved brain regions. Additionally, as we shall examine in more detail later on in this paper, neurons farther along in a sensory processing stream demonstrate increasingly advanced and abstract response properties. In this way, neurons in an anatomically 'higher' position on the neuroaxis also seem to possess 'higher' or more advanced response characteristics than neurons earlier or lower in the processing stream. All of these features suggest that consciousness might emerge at the highest levels of the neural hierarchy.

In spite of the intuitive appeal of emergentist accounts of consciousness, I will argue that the view that consciousness and mind emerge hierarchically at the 'summit' of the nervous system is a serious shortcoming of these theories. I will attempt to show that while it is true that all living things *are* hierarchically organized, the neural hierarchy in actuality does not have a 'top' or 'bottom' like a pyramid. Living things are structured as *nested* hierarchies, and I will argue that the neurobiological basis of consciousness can be understood as a *nested hierarchy of meaning and purpose*.

Finally, I will attempt to demonstrate that while radical emergence does not account for the non-reducibility of the mind to the brain, there are non-reducible features of mind that are explained by the inability to reduce the subject to the object, a feature of consciousness that can be understood as the natural result of brain evolution and normal neural functioning.

II: Emergence

The first concept that needs to be addressed in order to examine the claim that consciousness is an emergent feature of the brain is the idea of *emergence* itself (for reviews, see Beckermann *et al.*, 1992; Van Gulick, 2001 [this volume]). According to Jaegwon Kim, the concept of emergence holds that:

> ... although the fundamental entities of this world and their properties are material, when material processes reach a certain level of complexity, genuinely novel and unpredictable properties emerge, and that this process of emergence is cumulative, generating a hierarchy of increasingly more complex novel properties. Thus, emergentism presents the world not only as an evolutionary process but also as a *layered structure* — a hierarchically organized system of levels of properties, each level emergent from and dependent on the one below (Kim, 1992).

One popular example of a system with emergent features is water, for which it is noted that the properties of liquidity, wetness, and transparency do not apply to a single water molecule, but do apply to the aggregate 'water'. As noted by Kim, emergent properties are generally viewed as the result of *hierarchically* ordered entities (Allen and Starr, 1982; Ayala and Dobzhansky, 1974; Pattee, 1973; Salthe, 1985; Whyte *et al.*, 1969;). In accord with Kim's definition cited above, emergence in living things is said to occur as the result of a hierarchically organized system — the organism — where each level of complexity of the organism in theory produces *novel* emergent features from the levels below it.

For example, C. Lloyd Morgan, a leader of the school of emergentism, described the emergence of mental phenomenon in terms of a hierarchical biological model. Morgan viewed the mind as emerging at the 'peak of the pyramid' of the biological hierarchy:

> In the foregoing lecture the notion of a pyramid with ascending levels was put forward. Near its base is a swarm of atoms with relational structure and the quality we may call atomicity. Above this level, atoms combine to form new units, the distinguishing quality of which is molecularity; higher up, on one line of advance, are, let us say, crystals wherein atoms and molecules are grouped in new relations of which the expression is crystalline form; on another line of advance are organisms with a different kind of natural relations which give the quality of vitality; yet higher, a new kind of natural relatedness supervenes and to its expression the word 'mentality' may be applied (Morgan, 1923, p. 35).

Searle (1992) points out that there are at least two interpretations of the concept of emergence. The first he calls *emergence1*. This refers to the idea that a higher order feature of a system can be understood by a complete explication of the parts of a system and their interactions. A novel feature of a system produced in this fashion he terms a 'causally emergent system feature'. According to this definition of emergence, consciousness is a causally emergent feature of the nervous system in the same way that liquidity is an emergent feature of individual water molecules.

Searle distinguishes emergence1 from what he describes as a 'more adventurous conception' of emergence called *emergence2*. A feature of a system is considered emergent2 if it has causal powers that cannot be explained by the parts of a system and their interactions. According to Searle:

> If consciousness were emergent2, then consciousness could cause things that could not be explained by the causal behavior of the neurons. The naïve idea here is that consciousness gets squirted out by the behavior of the neurons in the brain, but once it has been squirted out, then it has a life of its own (Searle, 1992).

Van Gulick (2001 [this volume]) in a recent and comprehensive analysis of the topic of emergence, was able to discern ten varieties of the concept of emergence (and ten varieties of the converse idea of reduction; see below). One of Van Gulick's varieties of emergence he calls *radical kind emergence* that roughly corresponds to Searle's emergence2. According to Van Gulick, in radical kind emergence, 'The whole has features that are both 1. different in kind from those had by its parts and 2. of a kind whose nature and existence is not necessitated by the features of its parts, their mode of combination and the law-like regularities governing the features of its parts.' (Van Gulick, 2001 [this volume], p. 17). In this way, the emergent feature is said to be 'greater than the sum of its parts'.

However, as Van Gulick rightly points out, a theory that claims that consciousness is a radically emergent feature of the brain poses problems for a fully physicalistic account of the relationship between the brain and the mind:

> The notion that causal powers might exhibit radical kind emergence merits special attention since it poses perhaps the greatest threat to physicalism. If wholes or systems could have causal powers that were radically emergent from the powers of their parts in the sense that those system-level powers were not determined by the laws governing the powers of their parts, then that would seem to imply the existence of powers that could override or violate the laws governing the powers of the parts, i.e. genuine cases of what is called 'downward causation' ... in which the macro powers of the whole 'reach down' and alter the course of events at the micro level from what they would be if determined entirely by the properties and laws at the lower level (Van Gulick, 2001 [this volume], pp. 18–19).

For example, the psychologist Roger Sperry, who received the Nobel Prize for his research on patients with surgical division of their hemispheres, viewed the mind as an emergent2 or radically emergent feature of the brain. Like Morgan, Sperry viewed the mind-brain relationship in hierarchical terms and supposed that the neural elements of the brain combine in increasingly complex configurations until at the summit of organization, the mind emerges:

... consciousness was conceived to be a dynamic emergent of brain activity, neither identical with, nor reducible to, the neural events of which it is mainly composed. Further, consciousness was not conceived as an epiphenomenon, inner aspect, or other passive correlate of brain processing, but rather to be an active integral part of the cerebral process itself, exerting potent causal effects in the interplay of cerebral operations. In a position of top command at the highest levels in the hierarchy of brain organization, the subjective properties were seen to exert control over the biophysical and chemical activities at subordinate levels. It was described initially as a brain model that puts 'conscious mind back into the brain of objective science in a position of top command . . . a brain model in which conscious, mental, psychic forces are recognized to be the crowning achievement . . . of evolution' (Sperry, 1977; reprinted in Trevarthen, 1990, p. 382).

Sperry conceived of consciousness as a final and highest emergent feature of the neural hierarchy:

We do not look for conscious awareness in the nerve cells of the brain, nor in the molecules or atoms of brain processing. Along with the larger as well as lesser building blocks of brain function, these elements are common as well to unconscious, automatic and reflex activity. For the subjective qualities we look higher in the system at organizational properties that are select and special to operations at top levels of the brain hierarchy and which are seen to supersede in brain causation the powers of their neuronal, molecular, atomic, and subatomic infrastructure (Sperry, 1984, p. 671).

To summarize, Sperry argues that the mind emerges at the 'top levels of the brain hierarchy' where it occupies 'a position of top command'. Since Sperry's account posits that the mind is a radically emergent feature of the brain, according to Sperry, 'conscious phenomena are different from, more than, and not reducible to, neural events' (Sperry, 1977; reprinted in Trevarthen, 1990, p. 383; see also Sperry, 1966; 1984). Thus, Sperry's account does not provide a scientific account of how the mind can be reduced to the brain. This is the question I will consider next.

III: Reduction

In their book *The Life Science*, Medawar and Medawar (1977) claim that non-reducibility of a higher level to a lower level in a hierarchical system is the essential and defining characteristic of an emergent property. They define emergence as the 'philosophical doctrine opposed to *reducibility* which declares that in a *hierarchical* system each level may have properties and modes of behavior peculiar to itself and not fully explicable by analytic reduction.' According to the Medawars, if a property is emergent, it is *not* reducible to those parts. It follows that if a property in a hierarchy *is* reducible to the properties of things at lower levels of the hierarchy, it is not emergent. Their definition of an emergent feature corresponds to the characteristics of a radically emergent feature as defined by Van Gulick (2001 [this volume]) or an emergent2 property as defined by Searle (1992).

Searle (1992) argues that a satisfactory reduction involves a form of identity relation that he calls a 'nothing-but' relation: a property A is said to be reducible to property B if it can be shown that A 'is nothing but' B. In other words, we do not need to invoke a new or novel property to explain A over and above those principles by which we understand B. While Searle identified several subtypes of reduction, what he called *ontological reduction* is the form of reduction most relevant to emergence theory:

> The most important form of reduction is ontological reduction. It is the form in which objects of certain types can be shown to consist in nothing but objects of other types. For example, chairs are shown to be nothing but collections of molecules. This form is clearly important in the history of science. For example, material objects in general can be shown to be nothing but collections of molecules, genes can be shown to consist in nothing but DNA molecules (Searle, 1992, p. 113).

There are many examples of successful ontological reductions in neuroscience. Our understanding of the manner in which muscle movement can be reduced to the physiology of the nerve and muscle would be a simple example. Neurological processes that at one time were considered mysterious can now be explained in purely biological terms. We do not have a 'seizure-brain problem' because we understand how the observable epileptic fit can be reduced to abnormal electrical discharges of cortical neurons. However, consciousness has proved to be particularly resistant to a simple scientific reduction to the brain. If consciousness can be shown to be in fact be a radically emergent feature of the brain, this could in principle account for a failure, on the basis of neurological principles, to reduce consciousness to the brain. And this in turn would force one to endorse a form of dualism. I will suggest a way out of this dilemma, but first I will describe a third important aspect of emergence theory, *constraint*.

IV: Constraint

A third key concept of emergence theory is the notion of *constraint*. While the term emergence refers to the way that the parts in a hierarchy combine to form wholes at higher levels of a hierarchy, constraint is the process by which higher levels in the hierarchy impose control over the lower levels. Campbell (1974) coined the term *downward causation* to refer to the control that a higher level of a hierarchy exerts on its contributing parts. The biologist H.H. Pattee (1970; 1973) pointed out the importance of constraint in biological systems:

> If there is to be any theory of general biology, it must explain the origin and operation (including the reliability and persistence) of the hierarchical constraints which harness matter to perform coherent functions. This is not just the problem of why certain amino acids are strung together to catalyze a specific reaction. The problem is universal and characteristic of all living matter. It occurs at every level of biological organization, from the molecule to the brain. It is the central problem of the origin of life, when aggregations of matter obeying only elementary physical laws first began to constrain individual molecules to a functional, collective behavior. It is the central problem of development where collections of cells control the growth or genetic expression of individual cells. It is the central problem of biological evolution in which groups of cells form larger and larger organizations by generating hierarchical constraints on subgroups. It is the central problem of the brain where there appears to

be an unlimited possibility for new hierarchical levels of description. These are all problems of hierarchical organization. Theoretical biology must face this problem as fundamental, since hierarchical control is the essential and distinguishing characteristic of life (Pattee, 1970).

Hierarchical biological systems operate via constraint. The individual cells of the human body constrain the microscopic organelles of the cell to perform sub-cellular metabolic processes, and the organs of the body in turn constrain these cells to perform functions such as secretion or muscular contraction. The entire body of the person constrains the individual organs to breathe, digest, and perform the macroscopic functions necessary for life. For example, the mitochondria, the organelles responsible for generating energy from oxygen via cellular respiration, are a microscopic part of the cell that along with other cells make up the tissues that eventually give rise to the lung. The mitochondria contribute to the emergence of the lung at a higher level of the hierarchy of the body. The lung, at a higher level on the hierarchy, displays emergent features not possessed by mitochondria, for example breathing. If the lung did not breathe, the body would not have oxygen, and if we did not have oxygen, the mitochondria would not be able to carry on cellular respiration. The mitochondria contribute to the emergence of the lung, and the lung in turn constrains the mitochondria.

Note that in the above description of the operation of the lung, the type of emergence that is described is of the emergence1 variety. The operation of the higher levels of the hierarchy of the lung is clearly seen to be the result of the operation of the parts of the lung and their interactions. Therefore, it is not surprising or mysterious that the whole lung considered in its entirety can affect and constrain the individual cells and organelles within it. In contrast, in Sperry's view, the mind is an emergent2 or radically emergent feature of the brain. Thus although the mind emerges from the brain, and is not reducible to the brain, it can nonetheless *cause* material events to happen in the brain. In this way, Sperry suggested that '. . . subjective properties were seen to exert control over the biophysical and chemical activities at subordinate levels' (Sperry, 1977), and the immaterial mind possesses *causal* properties over the material brain and constrains it.

> The causal power attributed to the subjective properties is nothing mystical. It is seen to reside in the hierarchical organization of the nervous system combined with the universal power of any whole over its parts. Any system that coheres as a whole, acting, reacting, and interacting as a unit, has systemic organizational properties of the system as a whole that determine its behavior as an entity, and control thereby at the same time the course and fate of its components. The whole has properties as a system that are not reducible to the properties of the parts, and the properties at higher levels exert causal control over those at lower levels. In the case of brain function, the conscious properties of high-order brain activity determine the course of the neural events at lower levels (Sperry, 1977; reprinted in Trevarthen, 1990, p. 384).

However, as pointed out by Van Gulick (2001 [this volume]) and noted above, the idea that the mind is a radically emergent feature of the brain that is not reducible to the brain, and has causal powers over the brain, poses a real problem for a fully physical account of the brain/mind relationship. If we are to have a physicalistic account of the relationship between the brain and mind, we must have a model of the brain and mind that is hierarchical and allows for the creation of higher order mental features. But if the model is to avoid the dualism that is inherent in Sperry's conception, we should *not* posit the existence of radically emergent or emergent2 features. Finally, the model must also allow for the creation of higher order mental properties that provide the hierarchical constraint that guides abstract thought, perception, and volitional motor behaviours

V: Hierarchies and Emergence in the Brain

One problem with Sperry's account is that although he claimed that the mind emerges at the summit of a hierarchically ordered brain, there is in reality no 'summit of the brain'. There is no basis for the idea that the mind emerges, in Sperry's words, 'In a position of top command at the highest levels in the hierarchy of brain organization . . .' because in reality there is no 'top' or 'summit' in the brain. This can be seen simply in the visual system.

The visual hierarchy

Each cell of the retina responds to a particular area of the visual world called its *receptive field*. The receptive field of each retinal cell is quite small, and the brain must build up whole and unified mental representations from these discrete points of contact with the world. Hubel and Wiesel (1962; 1965; 1968; 1977; 1979; Hubel, 1988) in work that led to their receiving the Noble Prize in 1981, demonstrated that cells in cortical area V1, the primary visual cortical region, responded not to points of light but rather to thin bars of light. They called these cells *simple* cortical cells. They reasoned that a *single* neuron in the brain could respond to a *line* in the world if a line of adjacent firing cells — each responsive to an individual point of light — converged upon a single simple cortical cell further along the processing stream. That single cortical cell could 'add up' the points of light that each lower order cell had responded and in this fashion a *single* higher order cell could respond to a *line.*

Hubel and Wiesel also found that multiple simple cortical cells converged upon other single higher order cells further along the processing stream to create what they called *complex* cortical cells, and complex cells in turn converged upon single neurons to create *hypercomplex* cells with increasingly specific and complex response properties. The model of Hubel and Wiesel is hierarchical in that simple cells lower or earlier in the neural processing chain create cells of ever increasing complexity higher up on the neural hierarchy.

The progression from simple to complex to hypercomplex cells is an example of what Zeki (1993) terms 'topical convergence', the process whereby many lower order, simple visual cells simultaneously *converge* upon a smaller number of higher order complex cells. This hierarchical process ultimately produces advanced 'higher order' cells that possess amazingly specific response properties. For example, there are neurons in the inferior temporal cortex far 'downstream' from the simple cells found early in primary visual region that respond preferentially to highly specific and complex stimuli such as hands or faces. Some of these neurons are even responsive to a frontal view of a face while others react best to a side view. Neurons of this type led to the notion of the 'grandmother' cell, a cell so specific that it fires only to the face of ones own grandmother (Barlow, 1995). While cells of that degree of specificity do not exist, it is in general true that the farther along a sensory processing stream that one looks, the more specific a cell's response characteristics become.

However, in the process of creating higher order cells, the receptive fields of these cells increase in size. While a retinal cell early in the stream monitors a small and specific point in the visual field, a hypercomplex cell such as a face cell will react to a face that appears almost *anywhere* in the visual field. It follows that while the cells early in the visual stream with small receptive fields 'know' where each line of the face is, these early cells do not 'know' that a given line is part of a face. The face cells, on the other hand, 'know' there is a face, but due to the process of topical convergence, these cells don't know where the face is located in space. While it is true that cells of the brain project to successive levels in a hierarchical fashion in order to code for increasingly specific complex and abstract properties, the information coded by cells earlier in the process is not and cannot be lost in awareness.

An experiment by Movshon and co-workers (1985) also illustrates this problem. When certain shapes are viewed behind a small circular aperture, cells at the level of V1 would receive conflicting movement information. Under these conditions of viewing, adjacent sides of a horizontally moving diamond would appear to be travelling in different directions if visual information went no further than V1. The cells of V1, which possess small receptive fields, are able to encode the exact locus of a stimulus in space, but they are unable to discern the direction of movement of the entire object. In contrast, the cells of V5 are able to respond appropriately to the movement of the actual overall object, but these cells do not detect the exact topographical locus of the stimulus. Therefore, in order to know where the whole object moving in space is located, input from V1 and V5 is required and both V1 and V5 must make 'explicit' contributions to visual perception.

As Zeki observes, cells comprising *both* the early and the late stages of the visual processing stream must make a unique contribution to consciousness. In support of this position, Beckers and Zeki (1995) used Transcranial Magnetic Stimulation (TMS) to create a localized and reversable inactivation of the underlying brain. When the cells of area V1 of normal subjects are inactivated, conscious motion perception which requires an active V5 is still possible, but the ability to judge the orientation of stationary objects, a task that requires exquisite point localization, is lost. Under these conditions, the patient sees something move, but cannot identify the object. Beckers and Zeki conclude that V5 can make a conscious contribution to the perception of motion that is wholly independent of V1, the primary visual region.

Therefore, in contrast to Sperry's model, consciousness does not emerge at the 'top' of the visual hierarchy. Even though the brain seems to be creating 'pontifical neurons' at the 'top' of the visual hierarchy, the conscious mind is in reality 'spread out' across the activity of millions of neurons located in different regions and levels of the brain. A visual hierarchy exists, but many levels of the hierarchy make a contribution to conscious awareness. This is even truer when the brain must coordinate features of a stimulus from multiple sensory domains, such as an object having both visual and auditory features. When considering the perception of a honking red Corvette passing a brown barking dog, binding must occur in both visual and auditory centres that are located in even more widely distributed areas of the brain. The brain must bind the colour red and the honking horn to the car, and the colour brown and the barking to the dog it passes, and not the other way around. And both of these elements must be bound to each other within the overall perceptual field of awareness.

One way to explain binding (see for instance Newman, 1999; von der Malsburg, 1995) in the visual system is that the brain uses synchronized oscillations to unify perception. However, it seems likely that the brain uses many mechanisms to perceptually bind objects in awareness, including convergence, synchronized oscillations (Crick and Koch, 1990; Crick, 1990; Engel *et al.*, 1991; Gray and Singer, 1989; Gray *et al.*, 1989; 1992; Koenig and Engel, 1995; Newman, 1999) and re-entry (Edelman, 1989). For the present purpose, *asking which particular binding mechanism is used, and under what circumstances it occurs are not the critical questions*. Rather, from the standpoint of emergence theory, the point is that there is no 'top' of the visual pyramid where visual awareness may be said to radically emerge. All levels of the visual system lower and higher, contribute to consciousness.

The motor hierarchy

When we look for the 'top command' of the motor system, we are faced with the same dilemma. If you say to yourself 'I will now move my arm' and you do so, you will experience an 'inner I' as the source of that action. Suppose I, as your neurologist, search for the source of that 'will' somewhere in your brain. I will find that there is no central, integrated and unified physical locus that is the source of that action. Just as in the perceptual hierarchy that appears to lead to the 'grandmother cell' in the visual system, there is no singular 'top' of the motor hierarchy, no 'ghost in the machine' as Gilbert Ryle (1949) put it, that can serve as the source of our unified 'will'.

Consider the hierarchical organization of human speech. If we trace the control of our speech, at the lowest level of control, there are the neurons connected to the individual muscles of the lips tongue, face, etc. The control of these neurons can be ultimately traced to the highest levels of the central nervous system. If we look in the brain for the areas that control speech, we know first of all that the perisylvian speech regions of the left hemisphere are particularly important for the production of fluent speech, so that this area must be involved in the control of speech. We also know that the temperoparietal regions of the left hemisphere are critical for the knowledge of the words we use in speech, and that this region is also necessary for the comprehension of speech and for self-monitoring as we

WHY THE MIND IS NOT A RADICALLY EMERGENT FEATURE   133

listen to ourselves speak. We know the right hemisphere is important for creating emotional inflection in speech production, so include this area too. And if you gesture with your hands as you speak, we should include the motor areas of both sides of the brain as a part of the overall act of expressing oneself. It is clear that vast areas of the brain are involved in the complicated act of speech production, yet somehow when we 'will' to speak, the entire act is *unified* into a whole and coordinated behaviour. But there is no unified region or hierarchical level in the brain that in and of itself constrains our actions and intentions. Therefore, there does not appear to be any material 'top' to either the perceptual or motor hierarchy.

VI: Non-nested and Nested Hierarchies

A difficulty with Sperry's emergence account lies in the way that he viewed the hierarchy of the brain. Sperry viewed the brain and the mind as parts of a particular type of hierarchy known as a *non-nested hierarchy* (Allen and Starr, 1982; Salthe, 1985). A non-nested hierarchy has a pyramidal structure with a clear-cut top and bottom in which higher levels control the operation of lower levels. A classic example of a non-nested hierarchy is a military command in which a general at the top controls his lieutenants, who control their sergeants, and so on, down the chain of command until we finally reach the level of the individual troops. This would seem to be what Sperry had in mind when he spoke of a 'top command' that subordinated lower levels of a hierarchy. The non-nested hierarchy that Sperry envisioned is considered non-nested because while the successive levels of the hierarchy interact, each level of the hierarchy is physically independent from all higher and lower levels. The various levels of a non-nested hierarchy are not *composed* of each other.

An alternative framework for viewing the mind/brain relationship is another type of hierarchy known as a *compositional* or *nested hierarchy*. Allen and Starr (1982, p. 38) provide a definition of a nested hierarchy:

> A nested hierarchy is one where the holon at the apex of the hierarchy contains and is composed of all lower holons. The apical holon consists of the sum of the substance and the interactions of all its daughter holons and is, in that sense, derivable from them. Individuals are nested within populations, organs within organisms, tissues within organs, and tissues are composed of cells.

We refer to this type of hierarchy as nested because the elements comprising the lower levels of the hierarchy are physically combined or *nested* within higher levels to create increasingly complex wholes. The important distinction between non-nested and nested hierarchies is the relationship between the lower and higher levels of the hierarchy. A non-nested hierarchy has a clear top and bottom, and the control of the hierarchy comes from the top. A nested hierarchy has no top or bottom, and the control or constraint of the hierarchy is embodied within the entire hierarchical system. All living things, including us, are nested hierarchies. We are physically composed of minute organelles that are hierarchically organized to create a human being. In the hierarchy of a living person, it is the complete person who sits at the top of the hierarchy, and that person is not separate from the parts of which he or she is composed. Individual elements of the body that make up the person simultaneously contribute to the life of that person. The 'parts' of the person in this way are nested within the totality of the human being.

I suggest that the brain functions as a nested hierarchy, as do all biological systems, and that the proper model for the integration of the mind and the brain is that of a nested hierarchy. Consider again the face cells. The convergence of neural pathways makes possible a cell so specific that it will only fire to face. This process might lead one to think that a *single* 'grandmother' cell, at the top of the perceptual hierarchy, embodies the representation of an entire face in consciousness. But this would be a mistake. Rather, the entire nested system of the brain functions *interdependently* to create the visual image of the face. In the same way that mitochondria and the lung contribute to the life of the person, in the nested hierarchy of the mind, all the lower order elements — every line, shape, and patch of colour that make up our awareness of the face — continue to make a contribution to consciousness. In the perception of a speeding red Corvette, the roof and trunk of the car are composed of tens of thousands of individual line segments. These individual line segments of the car's outline then are combined into longer segments that produce the car's overall shape and form. The lower order features, for instance, the exact position of a small line sequence in space, emerge in awareness as 'part of' something else, such as the outline of the car. A short line segment is 'bound' to a longer segment to create the outline of the car, just as a small patch of red is 'bound' to a larger red patch that is part of the door. The redness of the Corvette is 'bound' to its shape which is bound to its movement which is bound to the honking horn, until all representations are 'bound together' to create the entire image. The colour, shape and movement of the car are nested together within the image of the car and this image in turn is nested within the entire scene.

To say an element is *bound* to another is simply another way of saying that they are represented in awareness *dependently* and are *nested together*. It follows from this framework that the extent to which lower order features are bound into a higher order feature is the extent to which the lower order features lose their independence from each other. For example, the neurons responding to the redness of the Corvette are tightly bound to each other. These neurons make their contributions to awareness as a nested whole. Every single neuron in a red patch of the car makes a unique contribution to consciousness, but it *becomes* a 'patch' of colour because all the neurons that represent redness make their contribution to awareness in an *entirely* dependent, bound, and nested fashion. The nested relationship can also be applied to the relationship between the redness and shape of the car. Colour and shape are represented in awareness as a nested totality. We do not experience the colour of the car independently from the experience of its shape. On the other hand, the experience of the dog that the car passes is bound to the car within the entire visual experience, but is not tightly bound to the colour or shape of the car. The higher level or complex neurons that code for car and dog as entities make greater *independent* contributions to conscious awareness within the nested hierarchy of consciousness than do the simple neurons that code for the specific colour or shape of the car.

VII: Emergence, Constraint and Reducibility in the Nested Brain

I would now like to return (in different order) to the three key elements that were addressed earlier — *emergence*, *constraint* and *reducibility* — and see how the nested model I propose deals with each of these features.

Emergence

According to a nested hierarchical model of the brain, consciousness does not mysteriously, or radically, emerge from the brain in some fashion that is independent from the brain. Indeed, higher order features of the brain are hierarchically created, but we can account for the creation of higher order mental features with reference to a nested system of lower and higher levels of organization. Therefore, in contrast to Sperry's emergence theory, this model of the brain accounts for both the creation of higher order mental features and mental unity without resorting to radical emergence or emergence2.

Constraint and perception

If there is no top or bottom to the nested hierarchy of the mind, what provides the top-most constraint that guides and controls our whole brain? In the hierarchy of our conscious awareness, it is meaning that provides the constraint that 'pulls' the mind together to form the 'inner I' of the self. Once again, analysis of visual processing can illustrate this point. Consider cyclopean perception and the 'mind's eye'. Sir Charles Sherrington, the father of modern neurophysiology, wondered how is it possible that there is a 'singleness' of normal binocular vision — known as the cyclopean eye — when either eye alone is able to generate a separate mental image? Sherrington noted:

> Our binocular visual field is shown by analysis, to presuppose outlook from the body by a single eye centred at a point in the midvertical of the forehead at the level of the root of his nose. It, unconsciously, takes for granted that its seeing is done by a cyclopean eye having a center of rotation at the point of intersection just mentioned (Sherrington, 1941, p. 277).

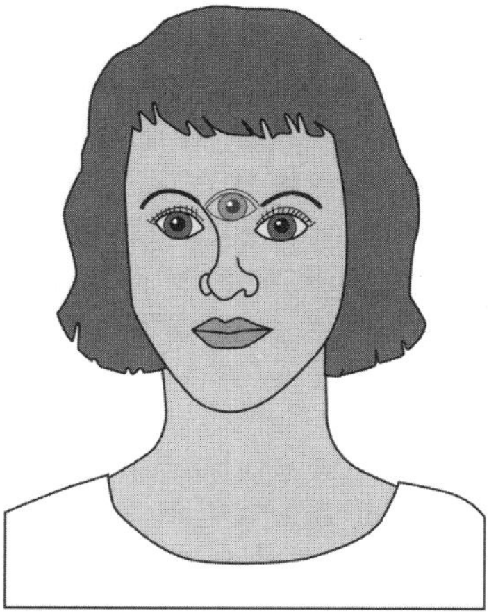

Simply put, the mind seems to have a visual synthesis point originating somewhere smack between our eyes and behind the top of our nose, as if there were a single cyclopean eye looking out from this point on the forehead (Figure 2). It is from this central — and single — vantagepoint that we experience the world visually as a coherent entity.

The problem is, needless to say, that we do not have such an eye located in the middle of the

Figure 2.
Although we have two eyes that possess overlapping visual fields, the brain creates a single 'mind's eye' that seems to be located at a point somewhere between and behind the actual eyes.

brain; rather, the brain works in a fashion such that we seem to have a central, cyclopean eye. In order to create cyclopean perception, the brain does not 'physically' merge the two visual images, one from each eye. Rather, the two images create a mental condition where their meanings are conjointly represented in awareness to produce a higher level of meaning. In the cyclopean eye, the higher level of meaning is that of a single image now seen in depth. It is the higher meaning of the combined images that produces the 'top-down' constraint upon the individual elements. This constraint or control of the whole upon the parts does not suggest the elimination of these individual parts from the mind because the image from each eye continues to make a contribution to the unified cyclopean eye. Instead constraint leads to the elimination of the independence of each part from each other when operating within the framework of the nested hierarchy of consciousness. By saying that these parts are represented dependently I also am saying that they are represented meaningfully in consciousness.

Constraint and volition

When the control of action or the nature of volition is considered as a nested hierarchy, it is purpose that provides the constraint and guiding force of the self. It is purpose that is the 'ghost in the machine'. Neurological theories usually do not make reference to purpose or 'free will'. Invoking purpose in an explanation of a neurological phenomenon is seen as a form of teleological thinking — an effort to invoke the end result of a process as its cause, which is a logical impossibility (Mayr, 1974; Searle, 1992). We cannot say, for example, that we possess the corneal reflex — the reflexive blink when something comes in contact with the cornea — in order to protect the eye. This form of teleological reasoning does not constitute a neurological explanation for the mechanism of the reflex. A neurological explanation of the corneal reflex entails a description of the nerves that provide sensation to the eye and the nerves that blink the eye, and the neurologist does not need to invoke the purpose of the reflex to explain its action. On the other hand, the corneal reflex does serve a purpose in that it protects the eye. All living things are hierarchically organized such that its parts are constrained to perform certain goal-directed actions that serve a purpose for the organism. In a nested hierarchy, constraint is the means through which organisms achieve their goals.

Another example of this relationship between constraint and purpose is demonstrated by the absorption of water by a plant via its roots. The roots as part of the plant are constrained to perform the actions necessary for this vital function. This does not mean that the plant 'knows' the purpose of this function or that the plant performs the actions of absorption 'in order to' bring water to its leaves. It would be teleological thinking to explain the mechanism of water absorption on the basis of the plant having the purpose of absorbing water. Similarly, a human being's beating heart, which promotes the circulation of blood throughout the body, has a structure and biological design that enables it to perform the functions necessary to maintain the life of the organism. But the heart does not purposefully set out to perform the pumping of blood any more than the plant purposefully sets out to absorb water. Nonetheless, the roots of the plant and the beating heart have within their structures the biological organization to perform the actions that ensure the organism's survival.

These examples demonstrate that it is difficult to describe biological functions that serve the purpose of ensuring the organism's survival in non-teleological terms. Theoretical biologist Ernst Mayr suggested that one approach might be to view biological systems as teleonomic, a word derived from the Greek word telos meaning 'goal' or 'endpoint.' According to Mayr 'a teleonomic process or behavior is one which owes its goal-directedness to the operation of a program (Mayr, 1974; 1982). Evolutionary processes produce living things that are teleonomic. Genetically programmed biological processes, such as the respiration of the mitochondria, are teleonomic. An organism can perform a process that is teleonomic in that the process serves an adaptive purpose for an organism, yet the organism does not have to anticipate or 'know' the end point of that process.

It is fairly clear that plants and beating hearts do not anticipate the consequences of their actions. But what about a cat stalking its prey? This example is also an instance of a teleonomic process. A biological program guides the cat's stalking behaviour, in this case we call it an instinct, and it has a clear end point which is to catch and eat its prey. According to Mayr's conception of teleonomy, does the beating heart have the same teleonomy as a cat chasing a mouse or the early man who made the first stone tool?

Mayr wished to regard teleonomic processes as those that are end-directed, but not deliberately or intentionally so. Teleonomic language, phrases such as 'the function of the heart's beating' or 'the heart beats in order to' are acceptable as long as one realizes the metaphorical use of such language. But if teleonomic processes are not necessarily deliberate, purposive, intentional acts, then the question remains as to whether there is a biological and even philosophical distinction between the beating heart and the thinking and purposive brain of the tool maker. And if there is a difference between the heart and the brain, how are we to describe the difference in scientific terms?

The answer is that while hierarchical constraint produces integrated teleonomic functions in all life, only certain teleonomic systems, events, acts, etc. are also purposeful. Some neurological phenomena are teleonomic yet entirely non-purposeful. The corneal reflex, which consists of the blink of the eye when it is touched, serves to protect the eye and is therefore a teleonomic behaviour. However, this reflex can be invoked in someone in coma and the reflex is largely involuntary. Therefore, while the corneal reflex is teleonomic in that it serves a purpose for the organism, it is not purposeful.

On the other hand, the act of speaking beautifully illustrates an enormous constraint of purpose over a nested system of parts. When we speak, we do not *will* a particular muscle of the mouth to move in a specific and conscious way. We do not consciously send a command to the tongue to move left and then right and so on. Indeed, we couldn't do this even if we wanted to. First of all, the average speaker has no idea which particular muscles to move when speaking. We are not consciously aware which muscles of the mouth, tongue, pharynx, larynx, etc. must be called into action to produce a certain sound. Additionally, even if we did know which muscles to move, we lack the fine coordination required to voluntarily move a given muscle 'just so' to produce a given speech sound. Finally, even if we were to possess the necessary knowledge and control to voluntarily move each muscle of speech into action, if we were required to do so when speaking, we would lose the overall integration of the speech act. We need to focus on the idea we wish to express, rather than on the way our lips and tongue are moving, or on the way a single fibre in the tongue is firing. The actions of the neurons at these lower levels of the hierarchy are *nested* within the higher levels of the hierarchy and the purpose of the act provides the *constraint* of higher levels upon lower levels of the motor hierarchy, and in this case the purpose of speaking is the communication of an idea. When we speak, it is the idea we wish to express, the *purpose* of our speaking, that sits at the highest level of the action hierarchy, and the *degree of constraint and purpose* distinguishes intentional from nonintentional teleonomic behaviours. Beating hearts do not act purposefully; toolmakers do. And the greater the constraint over nested parts within a hierarchy, the greater a behavior is purposive and therefore conscious.

Reducibility

Finally, what can we say about the reducibility of the mind to the brain? Can we perform an ontological reduction of mind to brain in the same way that axonal firing can be reduced to a knee jerk? While I have argued that radical emergence does not account for the ontological separation of consciousness and the brain, I agree that consciousness cannot be entirely reduced to the brain. Based upon other neurological considerations, I will argue that the reason the mind cannot be equated with the brain is the absolutely personal nature of meaning and purpose.

Many scholars have argued that the subjective personal and objective viewpoints on the brain are different in fundamental ways (Güzeldere, 1995; Velmans, 1991a,b; 1995; 1996; Globus, 1973; Metzinger, 1995; for a neurological perspective, see Feinberg, 1997). Kant (1781) insisted that the perception of an object is presupposed by an experiencing self, and it therefore follows that the self can never be an object unto itself. Schopenhauer also insisted that no thing could be simultaneously subject and object:

> Our knowledge, like our eye, only sees outwards, and not inwards, so that when the knower tries to turn itself inwards, in order to know itself, it looks into a total darkness, falls into a complete void. That the subject should become object for itself is the most monstrous contradiction ever thought of: for subject and object can only be thought one in relation to the other. This relation is their only mark, and when it is taken away the concept of subject and object is empty: if the subject is to become the object, it presupposes as object another subject — where is this to come from? (Arthur Schopenhauer, cited in Janaway, 1989).

In his classic paper 'What is it Like to Be a Bat?' Nagel also insisted upon the subjective nature of mental phenomena and he argued against simple reductionism and physicalism because 'every subjective phenomenon is essentially connected with a single point of view, and it seems inevitable that an objective, physical theory will abandon that point of view' (Nagel, 1974, p. 437).

A central theme of Nagel's argument is that the subjective–objective dichotomy lies at the heart of the mind–body problem, and that 'the subjectivity of consciousness is an irreducible feature of reality.' (Nagel, 1986, p. 7).

Searle has also argued that the mind is phenomenally irreducible to the brain. Searle pointed out that successful reductions in science aim to *remove* any subjective element in the analysis. A feature of the world which 'appears' a certain way in non-reduced form is ultimately reduced to its scientific 'reality'. However, when we consider the mind, Searle argued that '. . . we can't make that sort of appearance–reality distinction for consciousness because consciousness consists in the appearances themselves. *Where appearance is concerned we cannot make the appearance–reality distinction because the appearance is the reality*' (Searle, 1992, p. 122) This led Searle to suggest that 'the ontology of the mental is an irreducibly first-person ontology' (p. 95).

The question is: what is the neurological basis of the first-person ontology of consciousness? It is important to recognize that neural complexity, in and of itself, does not create meaning, purpose, or consciousness. Some very primitive reflexes involve highly complex neural states but do not require consciousness. Consider for example frogs, cats and dogs, whose spinal cords have been surgically separated from the brain. The hind limbs in these animals still will withdraw reflexively in response to certain stimuli even though the spinal cord connected to that limb is disconnected from the animal's brain. Chemical irritants, tickling, or noxious mechanical irritation would normally invoke pain, itch, or other sensations in the intact animal. But since the involved segments of the spinal cord that receive these stimuli are disconnected from the brain (specifically the part of the brain called the thalamus, necessary for conscious pain perception) the animal 'feels nothing'. Sherrington (1947) pointed out that these reflexes occurred without mental accompaniment. Without the central connections to the brain, reflexes can occur and be accompanied by quite complex neural activity, but not accompanied by consciousness. Indeed, some surprisingly complicated behaviours turn out to be highly automatic. For example, one can cut the spinal cord of a cat so that the portion of the spinal cord controlling the legs is disconnected from the brain. If the animal is put on a treadmill, it is able to walk with a rhythmic pattern typical of the intact animal. No consciousness, no volition, only an isolated spinal cord and its local circuits control the stepping pattern (Kandel *et al.*, 1999).

Another circumstance that demonstrates neural complexity in the absence of mind is the unfortunate patient who has had his or her spinal cord severed. If the spinal cord is severed completely, there is a physical separation of the nervous system above and below the lesion. Neural pathways which under normal circumstances convey pain impulses can travel up the spinal cord only part way before being prevented by the cut through the spinal cord from reaching the parts of the brain where conscious sensation occurs. Under these circumstances, if one were to apply a painful stimulus to the foot, such as forcefully squeezing the toe, the foot might still withdraw — reflexively — yet the patient will report he 'feels' nothing.

On the other hand, while some quite complex reflexes can occur *without* consciousness, some very simple organisms display behaviours that I would argue *do* involve meaning and consciousness. In the frog, a small, stationary stimulus situated directly in front of the animal evokes no response; it is for all intents and purposes invisible to the animal. But a small moving stimulus anywhere in the frog's visual field evokes an immediate dart of its tongue. Lettvin and co-workers (1959) recorded electrical activity from single fibres in the frog optic nerve during various stimulation procedures. They found one fibre population, called 'net convexity detectors', that responded when a small dark stimulus entered the receptive field, stopped, and moved about in a jerky fashion — just the kind of activity a mosquito might make if it were flying about in range of the frog's tongue. In fact, they found these fibres so exquisitely suited to detect a flying insect that they suggested they were best described as 'bug perceivers'.

For the bug perceivers of the frog brain to have 'meaning' to the frog, the frog's brain has done something quite remarkable: it has created 'an object' for the frog. The neural activity that creates the meaning 'bug' is certainly within the frog's brain. But the frog reacts to that neural activity as though it were not occurring within itself. The frog reacts to that neural activity *as if* it were in the world signifying fly as something that occurs outside of the frog's brain and being. The creation of outside objects is the fundamental starting point of all minds and the manner in which meaning is created.

Sherrington called this projection of sensation on the body into the world *projicience,* and he had an explanation for its evolutionary origins. Sherrington reasoned that it was the development of the distance receptors (nose, ear and eye), that enabled the registration of sensation from stimuli at a distance (smell, sound and light). In order for an animal to register a stimulus that really is 'out there in the world', the animal had to be wired to react to the stimulus as if it were external and not on the body where the sensory stimulation really is occuring (Sherrington, 1947, p. 324).

In this context, consider the process of tactually identifying an object, a process known in neurology texts as stereognosis. For a simple object such as a stone, recognition is almost immediate, yet no single point of the body feels the entire stone. Rather, the overall shape of the stone is extrapolated from the bending of the joints, its size by the distance between the fingers as they spread to cover the stone's surface and its texture by the smoothness or roughness on the skin tips. All these sensations which are actually occurring within the subject's brain are *perceived* as attributes of the object (Brain, 1951) When the sensations of the body are referred to the stone, these sensations have created a *meaning* for the subject.

Now here is the crux of the mind–body problem. The very point at which certain neural firings take on subjective meaning is also the point at which the subject's point of view of its own brain and the observer's point of view of that brain diverge. In the case of the frog, when the frog reacts to a fly, the frog's brain takes on a dual aspect. From the 'outside' or external point of view, the point of view of the neuroscientist who studies the frog, the frog's neurons are palpable, material, tangible objects; but from the frog's 'inner' point of view, those same neurons mean 'bug'. It follows from this example of the frog's brain that meaningful neural states will always entail two irreducible perspectives: the 'inside' subjective perspective and the 'outside' objective perspective. The experience that the brain creates is meaningful only to one 'I', its possessor. In this way, meaning and consciousness are irreducibly personal. When a neurological event carries meaning, the subjective aspects of the experience cannot be reduced to the objective neurological events.

Consider again the patient with the spinal cord injury. Suppose I, the neurologist, also acquired the same spinal injury, such that I, like my patient, had no sensation from the waist down. It would make little difference, from the standpoint of my consciousness, if I squeezed my patient's toe or my own. In either circumstance the withdrawal of the toe could occur but neither of us would 'feel' anything. If there is no 'first person' or 'I' associated with the response, there is no mind or subjective meaning involved with the reflex. With the subjective aspect of the response entirely removed, the withdrawal of the foot occurred without a subject. If I as the neurologist wish to analyse in this circumstance the reflex, I have no problem reducing the entire reflex to the neurons that created it. I can reduce the entire neurological event, beginning to end, to the firing of the neurons involved in the spinal reflex without recourse to a 'mind' or self, or an inner 'I'. However, if either of us feels the toe as it is squeezed, if there is an 'inner I' that has an experience that means, 'pin', we do have a problem reducing the neurological event to the brain. Once an event is meaningful, as was the case with the frog and the fly, the observer's point of view and the subject's point of view diverge.

Therefore, neurological considerations indicate that the non-reducibility of consciousness to the brain is not based upon the non-reducibility of emergent neurological features of the brain. Rather, the non-reducibility of consciousness to the brain is based upon the non-reducibility of the subject of experience to the observer of that subject.

This is why qualia, like meaning and purpose, have a fundamentally, irreducibly, first person ontology. Consider again squeezing the toe of a patient whom has spinal cord damage. When the spinal cord is intact, it should be obvious that the qualia of having one's toe squeezed — the pain involved in the experience — differs fundamentally depending upon whether one has the experience of the pain oneself, or observes it as a brain having the experience of pain. What is the importance of these differences for the neurologist? If one actually feels pain, one experiences the quale 'pain'. When the neurologist observes the brain experiencing 'pain' from the outside, he sees specific patterns of neural activity that can be accurately defined, but cannot see in the brain something neurological that is equivalent to the experience of 'pain'. When viewed from the outside therefore, the quale 'pain' really does exist materially. From the outside point of view, my patient's qualia are illusory. Qualia are personal and the relationship between a given brain and a given mind differs whether one is the person having that brain and that experience. From the observer's point of view, we cannot ultimately reduce the experience of 'pain' to the neural state that create it because there is nothing material from the outside perspective to reduce. There is no materiality to the experience 'pain' from the observer's point of view, because the experience of pain from the inner point of view only exists as neural activity from the outside perspective.

Furthermore, just as meaning has a personal ontology, individual purpose also possesses a first-person ontology and exists only from the 'inside' perspective of the self. From the observer's standpoint, we cannot locate in the brain an individual's purpose. The 'will' is not something that can be touched or pointed to. One can identify the pattern of neural firing that creates volitional action within the brain, but there is nothing about the pattern of a particular neuron's firing that distinguishes its firing as part of a willful action *per se*. The ontology of purpose and action, like the ontology of meaning, is irreducibly personal.

This does not mean that qualia don't exist, or that the mind is 'immaterial'. I do not aim to support any form of Cartesian dualism. To deny that feeling or qualia or consciousness exists is not only wrong, it avoids the question of what qualia really are. I only wish to point out that qualia, like meaning and purpose, only exist from the subjective point of view of the self.

What is it about the brain, what neurological factor or property accounts for the brain's capacity to create subjective neural activities? There is a simple fact about the brain that is often neglected that provides the answer to this question. The conscious brain has no sensation of itself. It has been known since the time of Aristotle that the brain is insensate (Clark and O'Malley, 1996, p. 9). For instance sticking a pin in the cortex itself evokes no pain that is referable to the brain itself. The brain has no sensory apparatus directed toward itself. As Globus puts it, the brain does not 'represent in any way its own structure to the subject' (Globus, 1973, p. 1129; see also Globus, 1976). There is no way that the subject can become aware of his own neurons 'from the inside'. They can be known only objectively from the 'outside'. We have already seen that there is no 'inner eye', no inner homunculus watching the brain itself, perceiving its own neurons, no 'brain-skin' which feels the neurosurgeon's knife. When I test a patients' pinprick sensitivity by applying a pin to their hand, and I ask them to localize where on the body the sensation is, I have never had a single one point to their head. Conscious neural activity refers to things, not to the brain itself. Conscious neural states are about things, not about the neurons themselves.

Many philosophers have argued that one of the essential characteristics of the mind is the property of intentionality. The use of the word intentionality dates to medieval times. The term derives from the Latin verb *intendo*, which means to 'point at' or 'extend toward'. Intentional phenomena are said to be about or of or directed at something. For example, beliefs are considered intentional because they are about a state of affairs. A fear is considered an intentional state because it is a fear of something. Perceptual states are considered intentional because if I experience an object in the world, whether I see, hear, touch or smell it, then I have a perception of an object. In a similar way, action states are considered intentional when they are directed at something in the world.

The seventeenth century philosopher and psychologist Franz Brentano, argued that intentionality is the defining feature of the mental and only mental phenomena possess the characteristic of intentionality. Consider how Searle spoke about intentional states:

> The second intractable feature of the mind is what philosophers and psychologists call 'intentionality', the feature by which our mental states are directed at, or about, or refer to, or are of objects and states of affairs in the world other than themselves (Searle, 1984, p. 14; see also Searle, 1983).

From the philosophical perspective, intentional mental states are directed at or *refer* to something other than themselves. On the basis of neurological considerations, I have argued that the origin of meaning and minds — from frog to man — are based on the fact that the conscious brain does not refer to itself. The similarity between the neurological perspective and the philosophical point of view is striking.

VIII: Conclusions

In this article I have tried to demonstrate that positing the radical emergence of the mind from the brain is unnecessary, and that with the appropriate neurological model and analysis, an entirely physicalistic account of the mind–brain relationship is possible. By its very nature, the brain functions in a fashion that produces irreducibly personal mental states, and the failure of mental states to be completely reduced to neurological states is not based on any variety of radical emergence of the sort posited by Sperry and others. Rather, from the standpoint of neurology, the irreducibility of mental states to the brain, to what extent such irreducibility exists, depends solely on the inability to reduce the subjective to the objective.

References

Allen, T.F.H. and Starr T.B. (1982), *Hierarchy: Perspectives for Ecological Complexity* (Chicago: University of Chicago Press).
Ayala, F.J., and Dobzhansky, T. (ed. 1974), *Studies in the Philosophy of Biology: Reduction and Related Problems* (London: Macmillan Press).
Barlow, H. (1995), 'The neuron doctrine in perception', in *The Cognitive Neurosciences*, ed. M.S. Gazzaniga (Cambridge, MA: MIT Press).
Beckermann, A., Flohr, .H and Kim, J. (ed. 1992), *Emergence or Reduction? Essays on the Prospects of Nonreductive Physicalism* (New York: Walter de Gruyter).
Beckers, G. and Zeki. S (1995), 'The consequences of inactivating areas V1 and V5 on visual motion perception', *Brain*, **118**, pp. 49–60.
Brain, W.R. (1951), *Mind, Perception and Science* (Oxford: Blackwell Scientific Publications).
Campbell, D.T. (1974), 'Downward causation in hierarchically organized biological systems', in Ayala & Dobzhansky (1974).
Clarke, E. and O'Malley, C.D. (1996), *The Human Brain and Spinal Cord.* 2nd edition (San Francisco, CA: Norman Publishing).
Crick, F.H.C. (1994), *The Astonishing Hypothesis* (New York: Basic Books).
Crick, F. and Koch, C (1990), 'Towards a neurobiological theory of consciousness', *Seminars in Neuroscience*, **2**, pp. 263–75.
Dennett, D.C., and Haugeland, J.C. (1987), 'Intentionality', in *The Oxford Companion To The Mind*, ed. R.L. Gregory (New York: Oxford University Press).
Edelman, G.M. (1989), *The Remembered Present: A Biological Theory of Consciousness* (New York: Basic Books).

Engel, A.K., König, P., Kreiter, A.K. and Singer, W. (1991),'Interhemispheric synchronization of oscillatory neuronal responses in cat visual cortex', *Science*, **252**, pp.1177–9.
Feinberg, T. E. (1997), 'The irreducible perspectives of consciousness', *Seminars in Neurology*, **17**, pp. 85–93.
Globus, G.G. (1973), 'Unexpected symmetries in the "World Knot"', *Science*, **180**, pp. 1129–36.
Globus, G.G. (1976), 'Mind, structure, and contradiction', in *Consciousness and the Brain: A Scientific and Philosophical Inquiry*, ed. G.G. Globus, G. Maxwell and I. Savodnik (New York: Plenum Press).
Gray, C.M. Engel, A.K., König, P. and Singer, W. (1992), 'Synchronization of oscillatory neuronal responses in cat striate cortex: Temporal properties', *Visual Neuroscience*, **8**, pp. 337–47.
Gray, C.M., König, P.A., Engel, A.K. and Singer, W. (1989), 'Oscillatory responses in cat visual cortex exhibit inter-columnar synchronization which reflects global stimulus properties', *Nature*, **338**, pp. 334–7.
Gray, C.M., and Singer, W. (1989), 'Stimulus-specific neuronal oscillations in orientation columns of cat visual cortex', *Proceedings of the National Academy of Science, USA*, **86**, pp. 1698–702.
Güzeldere, G. (1995), 'Problems of consciousness: A perspective on contemporary issues, current debates', *Journal of Consciousness Studies*, **2** (1), pp. 112–43.
Horgan, J. (1999), *The Undiscovered Mind* (New York: The Free Press).
Hubel, D.H. (1988), *Eye, Brain, and Vision* (New York: Scientific American Library).
Hubel, D.H., and Wiesel, T.N. (1962), 'Receptive fields, binocular interaction and functional architecture in the cat's visual cortex', *Journal of Physiology* (Lond.), **160**, pp. 106–54.
Hubel, D.H. and Wiesel, T.N. (1965), 'Receptive fields and functional architecture in two non striate visual areas (18 and 19) of the cat', *Journal of Neurophysiology*, **28**, pp. 289–99.
Hubel, D.H. and Wiesel.T.N. (1968), 'Receptive fields and functional architecture of monkey striate cortex', *Journal of Physiology* (Lond.), **195**, pp. 215–43.
Hubel, D.H., and Wiesel, T.N. (1977), 'The Ferrier Lecture: Functional architecture of macaque monkey visual cortex', *Proceedings of the Royal Society of London B.*, **198**, pp. 1–59.
Hubel, D.H., and Wiesel, T.N. (1979), 'Brain mechanisms of vision', *Scientific American*, **241** (3), pp. 150–62.
Janaway, C. (1989), *Self and World in Schopenhauer's Philosophy* (Oxford: Clarendon Press).
Kandel, E.R., Schwartz, J.H. and Jessell, T.M. (ed. 1999), *Principles of Neural Science* (Norwalk: Appleton & Lange).
Kant, I. (1781), *Critique of Pure Reason*.
Kim, J. (1992), 'Downward causation in emergentism and nonreductive physicalism', in Beckermann *et al.* (1992).
Konig, P. and Engel. A.K. (1995), 'Correlated firing in sensory-motor systems', *Current Opinion in Neurobiology*, **5**, pp. 519–51.
Lettvin, J.Y., Maturana, H.R., McCulloch, W.S. and Pitts, W.H. (1959), 'What the frog's eye tells the frog's brain', *Proceeding Institute of Radio Engineers*, **47**, pp. 1940–51. Reprinted in *The Embodiment of Mind*, ed. W.S. McCulloch (Cambridge, MA: Harvard University Press, 1965).
Mayr, E. (1974), 'Teleological and teleonomic: A new analysis', *Boston Studies in the Philosophical Science*, **14**, pp. 91–117.
Mayr, E. (1982), *The Growth of Biological Thought* (Cambridge, MA: Harvard University Press).
Medawar, P.B. and Medawar, J.S. (1977), *The Life Science: Current Ideas of Biology* (New York: Harper & Row).
Metzinger, T. (1995), 'The problem of consciousness', in *Conscious Experience*, ed. T. Metzinger, (Paderborn: Schöningh / Thorverton: Imprint Academic).
Morgan, C.L. (1923), *Emergent Evolution* (London: Williams & Norgate).
Movshon, J.A., Adelson, E.H., Gizzi, M.S. and Newsome, W.T. (1985), 'The analysis of moving visual pattern', in *Pattern Recognition Mechanisms*, ed. C. Chagas, R. Gattass and V. Gross (New York: Springer).
Nagel, T. (1974), 'What is it like to be a bat?', *Philosphical Review*, **83**, pp. 435–50.
Nagel, T. (1986), *The View from Nowhere* (New York: Oxford University Press).
Newman, J. (ed. 1999), *Conscious and Cognition*, **8** (2), Special issue on Temporal Binding and Consciousness (Orlando: Academic Press).
Pattee, H.H. (1970) 'The problem of biological hierarchy', in *Towards a Theoretical Biology 3*, ed. C.H. Waddington (Chicago, IL: Aldine).
Pattee, H.H. (ed. 1973), *Hierarchy Theory: The Challenge of Complex Systems* (New York: George Braziller).

Ryle, G. (1949), *The Concept of Mind* (London: Hutchinson and Company, Ltd.).
Salthe, S.N. (1985), *Evolving Hierarchical Systems: Their Structure and Representation* (New York: Columbia University Press).
Searle, J.R. (1983), *Intentionality* (New York: Cambridge University Press).
Searle, J.R. (1984), *Minds, Brains and Science* (Cambridge, MA: Harvard University Press).
Searle, J.R.(1992), *The Rediscovery of the Mind* (Cambridge, MA: MIT Press, Bradford Books).
Sherrington, C. (1941), *Man on His Nature* (New York: The MacMillan Company).
Sherrington, C. (1947), *The Integrative Action of the Nervous System* (New Haven, NJ: Yale University Press).
Sperry, R.W. (1966), 'Brain bisection and mechanisms of consciousness', in *Brain and Conscious Experience*, ed. J.C. Eccles (New York: Springer-Verlag).
Sperry, R.W. (1977), 'Forebrain commissurotomy and conscious awareness', *Journal of Medicine and Philosophy*, **2**, pp. 101–26; reprinted in Trevarthen (1990).
Sperry, R.W. (1984), 'Consciousness, personal identity and the divided brain', *Neuropsychologia*, **22**, pp. 661–73.
Trevarthen, C. (ed. 1990), *Brain Circuits and Functions of the Mind* (New York: Cambridge University Press).
Van Gulick, R. (2001). 'Reduction, emergence, and other recent options on the mind–body problem: A philosophic overview', *Journal of Consciousness Studies*, **8** (9–10), pp. 1–34.
Velmans, M. (1991a) 'Is human information processing conscious?', *Behavioral and Brain Sciences*, **14**, pp. 651–69.
Velmans, M. (1991b), 'Consciousness from a first-person perspective', *Behavioral and Brain Sciences*, **14**, pp. 702–26.
Velmans, M. (1995), 'The relation of consciousness to the material world', *Journal of Consciousness Studies*, **2** (3), pp. 255–65.
Velmans, M. (1996), 'What and where are conscious experience?' in *The Science of Consciousness*, ed. M. Velmans (New York: Routledge).
von der Malsburg, C. (1995), 'Binding in models of perception and brain function', *Current Opinion in Neurobiology*, **5**, pp. 520–6.
Whyte, L.L., Wilson, A.G. and Wilson, D. (1969), *Hierarchical Structures* (New York: American Elsevier).
Zeki, S.A (1993), *A Vision of the Brain* (Oxford: Blackwell Scientific Publications).

Anthony Freeman

God As An Emergent Property

Treating conscious states as emergent properties of brain states has religious implications. Emergence claims the neutral ground between substance dualism (perceived as hostile to science) and reductive physicalism (perceived as hostile to religion). This neutrality makes possible a theory of human experience that is religious, yet lies wholly within the natural order and open to scientific investigation. One attempt to explain the soul as an emergent property of brain states is studied and found wanting, because of a dogmatic assumption that God is 'beyond all material form'. Reflection on the central Christian claim that Jesus Christ was human and divine suggests the alternative view that God and the soul are both emergent properties. Unlike the philosopher's or physicist's remote and isolated 'first cause', this God is immediate and personal and social.

I: Introduction

It is often assumed by believers and atheists alike that materialist accounts of human consciousness must be antagonistic to a religious view of human life. This explains why Richard Swinburne, professor of 'the Philosophy of the Christian Religion' at Oxford University, has recently published a revised version of his staunchly dualistic study of the mind–body problem (Swinburne, 1986/1997); and why the unfashionable adherence of neuroscientist John Eccles to Cartesian dualism (e.g. Eccles, 1984) was widely attributed to his strong Catholic faith. On the other side, leading physicalists such as philosopher Daniel Dennett and biologist Richard Dawkins are pleased to believe that their researches support a god-free view of the world.

But we should not accept too quickly that substance dualism is necessary for religion. Not all mainstream neuroscientists are lacking in religious conviction. The late James Newman for instance, a neuropsychologist who directed much time and energy to understanding the thalamocortical mechanisms of consciousness, was a devout Episcopalian (Baars, 1999). And Keith Ward, Regius Professor of Divinity at Oxford and a close neighbour of both Swinburne and Dawkins, is comfortable with a view of the mind–brain in which 'conscious states will be emergent properties of the physical system' (Ward, 1998, p. 141). Emergence is a form of materialism, because it denies that a complex system or organism has any extra added ingredient over and above its physical parts; but it appeals to those who are uneasy with reductive physicalism, because also denies the claim that such a system is 'nothing but' the sum of its parts. On the contrary, emergent properties are claimed as novel features of a system as a whole that are not features of its constituent parts separately (see the survey of positions on emergence/reductionism in Van Gulick, 2001 [this volume]). The paradigm example of emergence is the liquidity of water, which is not a property of individual H_2O molecules (Searle, 1992).

John Searle calls liquidity a 'causally emergent' feature of the system, since it cannot be explained or predicted just from the physical structure of its parts, but it can be explained if account is also taken of the causal interactions between those parts. According to Searle, consciousness is best understood as a causally emergent feature of the brain. That is to say:

> The existence of consciousness can be explained by the causal interactions between elements of the brain at the micro-level, but consciousness cannot itself be deduced or calculated from the sheer physical structure of the neurons without some additional account of the causal relations between them (Searle, 1992, p. 112).

Searle distinguishes his own view of consciousness from what he calls 'a much more adventurous conception', where the emergent property is alleged to have causal powers that *cannot* be explained even when the causal interactions of neurons are taken into account. One proponent of this bolder approach is Michael Silberstein (1998; 2001 [this volume]), who says that emergent properties 'possess causal capacities that are not reducible to any of the causal capacities of the parts; and such properties are potentially not even reducible to any of the relations between the parts' (1998, p. 468). It is this more radical version of emergence that appeals to Keith Ward. I shall argue, against Ward, that Searle's more modest version of emergent consciousness is sufficient to sustain a worldview with a meaningful place for religion and the concept of God.

II: Emergence Constrained By Dogma

Ward is concerned — as I am — to reconcile the scientific and religious (more specifically the Christian) outlooks. He is attracted to emergence first of all because its firm basis in the material world is agreeable both to current science and to the Hebrew/Christian Bible. Emergence may still trail behind varieties of functionalism and other reductive models as the preferred scientific and philosophical approach to the physical basis of consciousness, but it is in the mainstream. Among others, Scott (1995), Searle (1992), Silberstein (1998), Sperry (1991) and Van Gulick (1992) have all appealed to emergence in some form or another in their work on the subject.

As for the Bible, it actually opens with a materialist view of human origins. Humankind is formed out of the earth: 'dust thou art and unto dust shalt thou return' (*Genesis* 3.19). Today the power of these words is lost because they are read through Greek-Christian spectacles. They are taken to refer to our physical bodies only: our bones will one day lie in the dust, but our real selves — our non-material souls — are destined to live with God in an eternal spiritual world.[1] But that is not what the words say, and it is not what they meant for the deeply religious society that produced them. In the ancient Hebrew understanding of human nature, a living person was a live body, animated by an impersonal 'breath of life', and a dead person was a dead body, lacking the breath of life, 'dust to dust'. In fact in the ancient Hebrew there is no word for 'body' in contrast to 'mind' or 'soul', in the Cartesian sense. Instead there are different words signifying the whole human person under different aspects, such as their vitality or their mortality (see Kliegel, 1998).[2] This holistic approach contrasted with that of the ancient Greeks, who did envisage a non-physical entity, the soul, that constituted the essence of a human being: a living person was an embodied soul, and a dead person was a disembodied soul.[3] Although biblical scholars today are inclined to make less of this contrast than was fashionable some years ago, Christianity is nonetheless heir to both these traditions. Ward deliberately aligns himself with this biblical view against the dualists:

> They [the dualists] think of God making a complete spiritual thing, with its own personality, and then having to attach it to some physical body.... Whereas the biblical account is that man is a truly physical entity, touched with God's spirit. It is *this holistic entity* that knows and thinks and decides.... So when we speak of the soul we speak of this physical entity in its capacity for a responsible relation to God . . . (Ward, 1998, p. 147; original emphasis).

Whatever he means by the phrase 'touched with God's spirit', by speaking of the soul as a physical entity Ward firmly excludes the idea that it implies either a Cartesian mind or a Platonic soul. And earlier he had emphasized that the 'mind and brain are not two quite different, disconnected realities. They form parts of one integrated whole, in which the brain is the way the soul appears to others' (p. 142).[4]

The second reason that emergence appeals to Ward is because it provides for a switch-over from 'bottom-up' to 'top-down' causality, and so allows for human purpose and freewill. One definition, for example, in addition to saying that emergent properties 'possess causal capacities that are not reducible to any of the

---

[1] As in the prayer used at the late night service of Compline, which asks that, 'when our bodies lie in the dust, our souls may live with [God].' For a profane example of the same view, consider the song: 'John Brown's body lies amouldering in the grave, but his soul goes marching on.'

[2] Familiar English translations of the Hebrew do not always help us to appreciate this point. For instance, the word *nephesh* is translated 'soul' over 400 times (against 'life' just over 100 times) in the King James version of the Old Testament. But the word 'soul' is quite misleading if it results in *nephesh* being taken as equivalent to 'mind' or the Cartesian *res cogitans*. In modern translations it is routinely translated 'life' or 'person', or indeed omitted altogether (e.g. 'my *nephesh* is in bitterness' might simply become 'I feel bitter').

[3] The New Testament usage is harder to determine, because although written in Greek its thought is steeped in the Hebrew Old Testament. Some scholars (e.g. Bultmann, 1955) argue that a word such as *psyche* (often translated soul, in accordance with its classical Greek meaning) is better thought of as equivalent to *nephesh* (see previous note).

[4] Ward seems here to use the terms consciousness, mind and soul, interchangeably. This is controversial, but in view of the variety of definitions for all three terms (and the added problem of assigning precise synonyms for each in other languages) his usage will be followed in the present paper unless stated otherwise.

causal capacities of the parts', also says that they 'exert causal influence on the parts of the system consistent with but distinct from the causal capacities of the parts themselves' (Silberstein, 1998, p. 468). This feature is crucially important to Ward, because it allows him to tell his religious constituency that epiphenomenalism is false and there is purpose in the system:

> The important point is that conscious states are not just thrown up as a sort of irrelevant by-product of brain-activity, which itself carries on in the same old mechanistic, predictable way. Rather, conscious states will be emergent properties of the physical system and as such they will modify radically the nature of the system. . . . Its laws will still, if you like, be physical, but they will not be derivable from the simple general principles of inorganic physics (Ward, 1998, p. 141).

So far Ward has said nothing that would appear out of the ordinary to the consciousness-studies community, and we may happily to go along with him on these first two points. It is with his third and final reason for embracing emergence — and the one that makes him adopt the most radical version of it — that we need to part company with him. Although he still wants to affirm the physical origins of human consciousness (or mind or soul), he also believes that it has the capacity 'to stand outside the physical processes that generate it' and to exist free-floating with no physical substrate. But here he goes much too far, because not even Silberstein's radical emergence can bear this frankly dualist interpretation. It is more like Hasker's ontological dualism (Hasker, 1999; see Van Gulick, 2001 [this issue]), an even more extreme version of emergentism, and one apparently driven in its turn by the perceived needs of conservative Christianity. In a move that is both questionable and — as will be argued in the present paper — unnecessary, Ward writes:

> The most important characteristic of the soul is its capacity for transcendence. It has the capacity to 'exist', to stand outside the physical processes that generate it, and of which it is part. . . . The material is limited to a particular location in space and time. It is contained by the location. But the soul by nature 'transcends'; it is oriented away from itself, to what is beyond itself (Ward, 1998, pp. 142–3).

There are at least two problems with this passage. First, the Cartesian assumption of nonlocal mind and localized matter is unwarranted. Whatever the final outcome of the debates over quantum mechanics and its interpretation,[5] it is hard to see a return to an understanding of physical entities that does not include some aspect of 'nonlocality', 'entanglement', 'holism' or whatever. Secondly, the claim that the soul has the capacity 'to stand outside the physical processes that generate it,' is ambiguous. It could mean that the soul does need a physical instantiation of some kind, but that it need not be the one that originally produced it. This sounds like the familiar computational-functionalist model of mind as a program that could in principle be stored in any number of physical systems,

---

[5] Nick Herbert's *Quantum Reality* (1985) is a sure-footed guide to the world of quantum mechanics for those with no mathematics. For QM in relation to different aspects of consciousness see, e.g., Stapp (1996) on the 'hard problem' of conscious experience; Clarke (1995) on nonlocality of mind; Silberstein (1998) on emergence and the mind–body problem; Hunt (2001 [this volume]) on 'submerged' consciousness.

providing they are appropriately related.[6] But functionalism is associated with a reductionist rather than an emergent approach and is rejected by Ward as taking 'too low a view of matter' (p. 146). An alternative reading would be that the soul, once it has emerged, can exist free-floating and requires no physical substrate. Ward confirms that he has this in mind when he likens the soul's ceasing to be dependent on the brain to a child's ceasing to be dependent on the womb (p. 152), and to a butterfly emerging from a chrysalis (p. 145). But how can consciousness or the soul be *independent* of physical brain processes 'of which it is a part'? The sense in which one physical object, such as a baby or a butterfly, 'emerges' from another (a womb or a chrysalis) is quite different from the sense in which a qualitatively new property of a whole system 'emerges' from its constituent parts.

The reason for this illogical move on Ward's part soon becomes clear. Despite his repeated insistence on its physicality, he claims that 'there is a very real sense in which the soul looks beyond the material world for its proper fulfilment. It looks to a presence and perfection and purpose beyond all material form — to the reality of God, its true source and goal' (p. 149). Now the cat — or the dogma — is out of the bag. Ward may be open to a contemporary philosophical and scientific view of human consciousness, but his resulting model of the soul has then to fit into a quite different framework. This is a framework dominated by a predetermined view of God as 'beyond all material form' and it requires a human soul to be likewise purified of its physicality. That dogmatic constraint leads Ward himself into exactly the kind of trouble encountered by the out-and-out dualists whose approach he has already rejected.

Having previously committed himself to the view that conscious states cannot exist apart from the brain, Ward now has to invoke an alleged 'subject of consciousness', that is not so constrained, in order to account for the free-floating soul that his doctrine of God requires. Occam is stropping his razor. And Ward struggles still. Reluctant even now to bring himself to a point so clearly at odds with all his previous argument, he appears at first to tie this new entity also to its physical roots:

> It is, therefore, not the conscious states that can exist separately, but the subject that knows and remembers these states and has certain dispositions and goals. This subject is acquainted with physical realities, and its goals are formed by and rooted in physical realities. It has its place in the physical world; it has a locatable viewpoint and field of action, in space (p. 148).

But finally he gives up the unequal struggle:

> Yet [this subject] is itself beyond every spatial and publicly observable description; it is the subject that is never an object, even to itself; the agent that is never itself an event in the world (p. 148).

Ward is forced into this untenable position by his belief that the soul, although physical in origin, 'looks *beyond the material world* for its proper fulfilment. It looks to a presence and perfection and purpose *beyond all material form* — to the

---

[6] On this view, according to Ned Block's (1980) facetious suggestion, a single mind could even be instantiated in a physical substrate consisting of the entire Chinese nation!

reality of God, its true source and goal' (p. 149, emphasis added). In other words, Ward fails in his quest to reconcile religion with a contemporary philosophical and scientific view of human consciousness, because of a predetermined view of God as 'beyond all material form'. That dogmatic constraint forces him to regard emergence, not as the permanent relationship between the mental and physical, but as a temporary transition from a physical to a non-physical mode of being. What he needs is a more thorough-going application of the emergence approach, one that includes God as well as human consciousness, to avoid this dislocation in his account.

In the remainder of this paper I shall argue that Ward's dualism, and the view of God that gave rise to it, can be avoided by applying Searle's version of emergence to an ancient Christian formula concerned with the humanity and divinity of Christ.

III: God As 'Human-shaped'

Christianity is heir to both the holism of the Hebrews and the dualism of the Greeks, and its most formative period — the fourth- and fifth-centuries — saw Greek dualism in the ascendant. The classic Christian debates concerning humankind and God focused on the person of Jesus Christ, understood primarily as the one whose function was to save humanity from the consequences of sin. Three apparently incompatible things were felt to be necessary in this saviour: he had to be fully human, and he had to be fully divine, and he had to be a fully-integrated person as well. Classical christology (as the doctrine of the person of Christ is known) ultimately failed because any two of these conditions were only ever achieved at the expense of the third.[7] A number of models were invoked to explain how these three conditions might be met.

- One viewed Jesus as a fully human person inspired by God's spirit. This safeguarded his humanity, but was felt to compromise his fully divine status.

- An alternative suggestion was that in Christ the divine Word or 'Logos' (a key concept among the Stoics) replaced the human soul altogether as the animator of the body. This secured the divinity, but omitting the soul altogether was held to compromise Christ's full humanity.

- A third possibility, designed to overcome the shortcomings of the other two, was that in Christ the Logos was in the same relationship to the *whole human person* as the soul was to the body in any human person. This arrangement safeguarded both the divinity and the humanity, but it called into question the integrity of the person of Jesus. How could there be two controlling principles — the divine Logos and a human soul — in one body?

The final compromise, contained in the Chalcedonian Definition of Faith, was in fact based on this third model, but with the human soul regarded as so

---

[7] For a lucid account of the logic of doctrinal development at this time, see Wiles (1967).

subservient to the divine Logos that in practice it boiled down to the second option.[8] This third option became the official teaching and was enshrined in a number of ancient formulae, including the so-called 'Athanasian' Creed.[9] This states, among other things, that Christ is:

> Perfect God, and Perfect Man: of a rational soul and human flesh subsisting; ...
> Who although he be God and Man: yet he is not two, but one Christ; ...
> *For as the rational soul and flesh is one man: so God and Man is one Christ.*

Now let us feed back into this formula John Searle's non-dualist 'emergence' model of the mind–body relation. On this view, remember, the conscious mind has its origins in the physical brain, but is not simply the same thing as the brain. Having emerged from the physical body, but without any added ingredients such as Descartes' 'mind stuff', it exhibits new features over and above the sum of its parts. It takes on an existence of its own, which is more than just the subjective experience of the person concerned, and it has a legitimate place in the external world of bodies and events. But it cannot be altogether divorced from its physical substrate.

What happens when we apply this model of human consciousness to the formula from the Athanasian Creed? Simply this: *the divine element in Christ is now to be understood as an emergent property.* That is to say: just as Christ's human mind — and indeed any human mind — arose from the complex physiology of his body, especially his brain and nervous system, so his divinity arose from the complex system which was his total humanity, body/mind/soul/consciousness. To repeat: just as the mind or soul is not an added ingredient to the human body, but an integral emergent property of it, so Christ's divinity is not an added ingredient to his human person, but an integral emergent property of it. And the Christian tradition says that what is true of Christ's divinity is true of God absolutely.

Searle uses the formula 'caused by and realized in' to explain the relation between an emergent property of a system and the lower-level elements that make up the system. Liquidity is a higher-level property caused by and realized in H_2O molecules; human consciousness is a higher-level property caused by and realized in the physical structure of the brain and nervous system. By extension I am saying that 'God-consciousness' — or simply 'God' — is a higher-level property still, caused by and realized in the physical-and-mental-totality of human beings.

IV: A Major Objection

The usual immediate objection to this claim is that God is by definition the creator and therefore prior to any created thing; that God is by definition

---

[8] The key passage reads: 'This one and the same Jesus Christ, the only-begotten Son [of God] must be confessed to be in two natures, unconfusedly, immutably, indivisibly, inseparably [united], and that without the distinction of natures being taken away by such union, but rather the peculiar property of each nature being preserved and being united in one Person and subsistence, not separated or divided into two persons, but one and the same Son and only-begotten, God the Word, our Lord Jesus Christ.'

[9] Dating probably from the second half of the fifth century, a hundred years after the death of Athanasius, who was active at the Council of Nicaea.

un-caused and non-material and therefore cannot be 'caused by and realized in' human minds-and-bodies. There is an argument, going back at least to Thomas Aquinas in the thirteenth century — and based on ancient ideas of Aristotle — which claims that our contingent world (i.e. a world consisting of cause and effect) can only be explained by some original uncaused (or 'necessary') reality, namely God. A related argument, working not from the mere fact of the universe but from its apparent order and purpose, deduces that the world must have had a designer, namely God. Such attempts to bridge the gap from the observed world to the 'mind of God' are still popular, but they are misplaced. When a human being was thought of as composed of a mind temporarily attached to a body, it made sense to think of God as a free-floating Mind not attached to a body. But we no longer talk like that about our own minds and neither should we talk like that about God. But physicists still do it. Even Stephen Hawking, who does not believe in such a God, uses the phrase 'mind of God' in the final sentence of *A Brief History of Time* (1988). And Paul Davies quotes this in a book actually titled *The Mind of God* (1992), in which he appears to endorse a form of deism (the view that God is a mind behind the universe, but makes no continuing intervention in it), although he too rejects the Christian religion in its fulness. But these kinds of argument for this kind of God cannot bear the weight put upon them, as David Hume (1779) and others have long since shown.

Among theologians, whose primary concern with God is religious rather than cosmological, many have taken the lesson to heart. For instance, existentialists like Rudolph Bultmann and John Macquarrie have argued strongly against any teaching that might appear to 'objectify' God as creator. In existentialist philosophy, 'Being' is not itself 'a being' or an object, but rather that which 'lets be' and makes it possible for all individual beings or objects to exist. As existentialist theologians equate God with Being, they reject 'every sense in which God can be understood as an entity which can be objectivised . . .' (Bultmann, 1955, p. 287.) In similar vein, Macquarrie asks whether one ought to say that God exists and he concludes that

> while to say 'God exists' is strictly inaccurate and may be misleading if it makes us think of him as *some* being or other, yet it is more appropriate to say 'God exists' than 'God does not exist', since God's letting-be is prior to and the condition of any particular being (Macquarrie, 1966, p. 108).

In other words, he is saying that objective existence is altogether an inappropriate category to use of God (see further in Freeman, 1993).

It may now be further objected that by setting aside the mythical concept of a creator God we are redefining the term in an arbitrary fashion. This is not so. The seeds of our human-oriented approach to the concept of God may be found in the belief, common to Christianity and Judaism, that humankind was made in God's image, an image that early Christians believed had been renewed and perfected in the person of Jesus Christ. They of course did believe in God as an independent 'supernatural' and 'eternal' being, but the pattern of theology that they laid down — with Christ at the centre — makes it possible to draw on their insights and yet to shift from that preconceived idea of God to a new one derived from the human end of things. The crucial texts are in Saint John: 'No one has ever seen God. It is God the only Son [i.e. Jesus Christ] . . . who has made him known' (*John* 1.18), and Saint Paul: 'God was in Christ' (*2 Cor.* 5.19) and 'In [Christ] all the fullness of God was pleased to dwell' (*Col.* 1.19). So the New Testament not only prepares its readers, it positively requires them, to develop an understanding of God that is 'Christ shaped' and therefore 'human shaped'. Christians claim no *a priori* knowledge of God, but only what they know through Christ: what God-in-Christ is, God is. And if that turns out to be not a supernatural agent external to humanity, but an emergent property of human being itself, then so be it.

V: An 'Open' Understanding of Human Being

Another objection to the emergence view of God, this time from the religious side, might be that it breaks down the absolute boundary between the divine and human. To this it may be replied that some such breakdown is necessary.[10] The tyranny of pre-existing concepts of human nature and divine nature, each defined by certain fixed and incompatible characteristics, lay at the heart of the problem faced by Chalcedonian-style christology.[11] On this understanding, to be both God and man would mean being simultaneously omniscient and ignorant, immortal and mortal, all-powerful and weak, etc. Such an approach was bound to produce a circle that could not be squared. But the concept of a universal and eternally fixed human nature is not the only way to think about human being. Modern thinking, influenced Kant's 'shift to the subject', has moved away from the earlier idea of the divine Logos taking on a kind of impersonal 'humanity' and put the emphasis firmly on the individual humanity of Jesus of Nazareth.

One widely respected scholar who grasped the significance of Chalcedon's deficient concept of human nature was the catholic Karl Rahner, who died in 1984. He took as his starting point the conviction that *openness* is the fundamental human characteristic. 'Man is spirit,' he wrote, 'that is, he lives his life in a perpetual reaching out to the Absolute, in openness to God' (Rahner, 1969, p. 66). From this beginning, although he still used the old categories of the Chalcedonian Definition, Rahner was able to avoid the problems raised when christology is treated as an attempt to put together two quite different entities — humanity and divinity — in one person. For him, humanity in its openness was sufficiently like God (bearing the image of God) to make the conjunction possible. As he put it, somewhat scornfully:

> Only someone who forgets that the essence of man is to be unbounded . . . can suppose that it is impossible for there to be a man, who, precisely by being man in the fullest sense (which we never attain), is God's existence in the world (Rahner, 1961, p.184).

---

[10] The absolute barrier owes more to the Neoplatonism of the early Christian theologians than to the demands of religion. Within Christianity itself, the belief that Christ himself was both human and divine makes some form of compatability essential. In some other religious and spiritual traditions, the absolute barrier has never been acknowledged in the first place.

[11] For an alternative reading of Chalcedon, using complementarity arguments derived from wave/particle duality in quantum physics, see Reich (1990); cf. Polkinghorne (1998).

This puts an entirely different complexion on things. Divinity is not now seen as something over against humanity, different from it, incompatible with it, but quite the opposite. To be human 'in the fullest sense' is itself to be 'God's existence in the world'. Here is a theological approach much more conducive to 'emergence' than anything we have seen hitherto.

In fact the ideas were not new when Rahner put them forward. Similar views had been published nearly a century and a half earlier by Friedrich Schleiermacher (1768–1834), the protestant scholar who bears the honorific title 'father of modern theology'. Schleiermacher spoke of a quality — which he claimed must be present potentially in all human minds — that he called 'God-consciousness'. Any given person's degree of religious awareness was, in his view, a measure of how far this *potential* for God-consciousness had become *actual* in that person; and it was this quality that Schleiermacher said Jesus must have possessed to a hitherto unknown extent. Schleiermacher felt he had here a theological key to unlock the door barring the way between humanity and divinity. On the one hand, Christ was the final stage in human evolution, so that he could call him 'the one in whom the creation of human nature, which up to this point had existed only in a provisional state, was perfected' (Schleiermacher, 1989, p. 374). But this unique degree of God-consciousness resulted in something more. While remaining beyond any question a state of *human* perfection, its being a state of human *perfection* gave it an altogether new — indeed a divine — dimension. Thus we find Schleiermacher writing:

> The Redeemer *is like* all men in virtue of the identity of his human nature, and *distinguished* from all by the constant potency of his God-consciousness, which was a veritable existence of God in him (p. 385).

The 'like' in this quotation is more significant than the 'distinguished'. Christ's human nature is like ours *absolutely*; the constancy of his God-consciousness distinguishes him from us *not* absolutely, but only *by being the first*. Schleiermacher makes this clear when he writes, 'As certainly as Christ was a man, there must reside in human nature the possibility of taking up the divine into itself, just as did happen in Christ' (p. 64).

As it stands, the reference to the divine in this last quotation is ambiguous. It could be interpreted in a dualistic way to mean that the divine is something external that needs 'taking up . . . into' the human (rather as a sponge takes up water). But read alongside the earlier references it may also be construed in an evolutionary way, as meaning that the break-through to 'the veritable existence of God in [Christ]' is a stage — albeit an extraordinary and at this point unique stage — of the natural process of development, with the emergence at key points of new levels of existence. This interpretation becomes imperative when we continue with the next sentence so that the full quotation reads:

> As certainly as Christ was a man, there must reside in human nature the possibility of taking up the divine into itself, just as did happen in Christ. So the idea that the divine revelation in Christ must be something in this respect supernatural will simply not stand the test (p. 64).

Schleiermacher clearly saw no conflict between 'perfect human nature' and 'the taking up of the divine', and there seems to be a good match between emergentist philosophy and Schleiermacher's theology. This is an encouragement to keep an open mind about the nature of God, and to work towards an understanding of it on the basis of St. John's belief — and Schleiermacher's and Rahner's — that it is by looking at Christ's humanity, and therefore all humanity, that we shall learn what God is like.

VI: The Interpersonal Emergence of God

The ideas of Schleiermacher and Rahner sit comfortably with the emergence approach we have been considering. The difference between their handling of the God/man relation and that of Keith Ward is clear. He took a late-twentieth-century view of the human mind/soul as an emergent property and tried to match it to concept of God that belonged to a quite different and fundamentally dualistic attitude to the mind–body question. It was a concept of God, moreover, that related more to the questions asked by philosophers and physicists about the ultimate cause and origin of the universe, than to the God of religious experience. The view we have now been considering is an essentially religious approach to God. Unlike the philosopher's or physicist's remote and isolated 'first cause', this God is immediate and personal — and also social. The complex human organism that provides the substrate for the emergent divinity need not be limited to individuals, but may be thought of interpersonally.

It was a weakness in Schleiermacher's theory that he failed to bridge adequately the gap between the God-consciousness of the Redeemer (as he liked to call Jesus) and the God-consciousness subjectively experienced by members of the Christian community centuries later. There was for him an unresolved tension between his commitment to the central importance of the individual subject and his awareness of the importance of the social dimension of religion. In view of the matching which we have suggested between Schleiermacher's theology and John Searle's philosophy, there may be a way of developing an interpersonal dimension by invoking a recent critique of Searle's work, precisely on the grounds that it is too individualistic (Núñez, 1995). Rafael Núñez has developed Searle's ideas in a more social direction (taking account of what he calls 'supra-individual biological processes' in addition to Searle's individually conceived 'biological naturalism'). These further suggestions may be applied to strengthen the link between Christ and the contemporary Church, and so between God as emergent in one unique human being and from within a whole community.

Searle insists that mental events are as much part of our biological natural history as digestion or enzyme excretion, but he limits this to the individual body. Núñez agrees that mental events are inseparable from biological ones, but points out that to explain certain *cultural* features of consciousness we need to take into account the social interaction of individuals. He uses the phenomenon of regional accents as an analogy:

> We all agree that one needs a brain in order to 'produce' a specific accent. Nevertheless in order to explain, say, your particular accent, it is not enough to study your brain as an isolated organ. It is inadequate to propose an explanation of your accent *only* in terms of what happens with your neurones [. . . and the rest of your body's system of speech-production]. In fact, nothing in your 'wiring' or in your genetic apparatus says that you should have a mid-western accent or a Welsh one. Certainly your brain is necessary for your accent to occur, but as an isolated organ does not determine it. Accents are modes of 'producing noises' during speech whose distinction is made at collective levels (at supra-individual ones), and whose frequency and distribution are neither determined genetically, nor at random. Nonetheless, an accent is still a living phenomenon. Accents have to do with biological processes that are irreducible to one individual's biology. . . . I claim that when it comes to the study of mind, consciousness and cognition (especially high level cognition), we face a situation analogous to this one (Núñez, 1995, pp. 162–3).

If Núñez is right, then we can and ought to include Schleiermacher's God-consciousness among those mental states which come about in the context of a supra-individual biological process, in particular the cultural interaction within the community of faith. This makes room for another historically dominant Christian insight, that the community (the Church), rather than the individual, is the locus of God's abiding presence.

VII: Looking Forward

Although we have used here examples from the Christian religion, the view that has been argued for will find points of contact in other spiritual traditions. Indeed, if its central thesis is right, and divinity is to be understood as emergent from and present within human life, then any attempt to restrict God to one particular history or tradition becomes ridiculous. Rather, by conceiving of God as organically one with humanity, both emerging from it and also inspiring it to greater and 'higher' achievements, this approach may provide a positive channel of communication not only between the various religions, but also between them and other areas of human endeavour as diverse as music, literature and current science.

There may be some who ask, Why bother? Why not let religion die a decent death and leave God-language in the past where it belongs. For the present writer there are two answers to this. The first and general point is that religion historically has provided the vocabulary and context for humans to explore their deepest needs and highest aspirations. That accumulation of wisdom and insight deserves to be studied and understood. And secondly, more specifically, the field of consciousness studies should not be fenced in with artificial boundaries, but continue to be explored until its natural limits become apparent.

References

Baars, B. (1999), 'James Newman: Obituary notice', *Journal of Consciousness Studies*, **6** (6–7), p. 201.
Block, N. (1980 [1978]), 'Troubles with functionalism', in *Readings in Philosophy of Psychology* (Cambridge, MA: Harvard University Press).

Bultmann, R. (1955), *Essays: Theological and Philosophical* (London: SCM Press).
Clarke, C.J.S. (1995), The nonlocality of mind', *Journal of Consciousness Studies*, **2** (3), pp. 231–40.
Davies, P. (1992), *The Mind of God* (London: Simon & Schuster).
Eccles, J.C. (1984), *The Human Mystery* (London: Routledge and Kegan Paul).
Freeman, A. (1993), *God In Us* (London: SCM Press; 2nd ed., Thorverton: Imprint Academic, 2001).
Hasker, W. (1999), *The Emergent Self* (Ithaca, NY: Cornell University Press).
Hawking, S. (1988), *A Brief History of Time* (London: Bantam)
Herbert, N. (1985), *Quantum Reality: Beyond the New Physics* (New York: Doubleday).
Hume, D. (1779), *Dialogues Concerning Natural Religion*, ed. M. Bell (Harmondsworth: Penguin Books, 1990).
Hunt, H.T. (2001), 'Some perils of quantum consciousness: Epistemological pan-experientialism and the emergence–submergence of consciousness', *Journal of Consciousness Studies*, **8** (9–10), pp. 35–45.
Kliegel, M. (1998), 'How important is an appropriate anthropological foundation of an ethic of human brain research? Philosophic-theological remarks', *Consciousness Research Abstracts* (a service from the *Journal of Consciousness Studies*), Abstract no. 472.
Macquarrie, J. (1966), *Principles of Christian Theology* (London: SCM Press).
Núñez, R. (1995), 'What brain for God's eye? Biological naturalism, ontological objectivism and Searle', *Journal of Consciousness Studies*, **2** (2), pp. 149–66.
Polkinghorne, J. (1998), *Belief in God in an Age of Science* (Yale University Press).
Rahner, K (1961), *Theological Investigations*, Vol. 1 (London: Darton, Longman & Todd).
Rahner, K (1969), *Hearers of the Word* (Herder & Herder).
Reich, K.H. (1990), 'The Chalcedonian Definition: An example of the difficulties and the usefulness of thinking in terms of complementarity?', *Journal of Psychology and Theology*, **18** (2), pp. 148–57.
Schleiermacher, F. (1989), *The Christian Faith* (Edinburgh: T & T Clark).
Scott, A. (1995), *Stairway to the Mind* (New York & Berlin: Springer).
Searle, J.R. (1992), *The Rediscovery of the Mind* (Cambridge, MA: MIT Press).
Silberstein, M. (1998), 'Emergence and the mind–body problem', *Journal of Consciousness Studies*, **5** (4), pp. 464–82.
Silberstein, M. (2001), 'Converging on emergence: Consciousness, causation and explanation', *Journal of Consciousness Studies*, **8** (9–10), pp. 61–98.
Sperry, R.W. (1991), 'In defense of materialism and emergent interaction', *Journal of Mind and Behavior*, **12**, pp. 221–45.
Stapp, H. (1996), 'The hard problem: A quantum approach', *Journal of Consciousness Studies*, **3** (3), pp. 194–210.
Swinburne, R. (1986/1997), *The Evolution of the Soul* (Oxford: Oxford University Press).
Van Gulick, R. (1992), 'Nonreductive materialism and intertheoretical constraint', in *Emergence or Reduction? Essays on the Prospects for Nonreductive Physicalism*, ed. A. Beckermann, H. Flohr and J. Kim (Berlin: DeGruyter).
Van Gulick, R. (2001), 'Reduction, emergence and other recent options on the mind–body problem: A philosophic overview', *Journal of Consciousness Studies*, **8** (9–10), pp. 1–34.
Ward, K. (1998), *In Defence of the Soul* (Oxford: Oneworld).
Wiles, M. (1967), *The Making of Christian Doctrine: A Study in the Principles of Early Doctrinal Development* (Cambridge: Cambridge University Press).
Young, F. (1983), *From Nicaea to Chalcedon: A Guide to the Literature and its Background* (London: SCM Press).

Alwyn Scott

We Could Be Siblings Yet

Reflections on Huston Smith's 'Why Religion Matters'

In the course of a distinguished academic career, Huston Smith passed some fifteen years in the Department of Humanities at MIT where he had abundant opportunity to test his lifelong beliefs against the doctrines of the scientific community. The loving fruit of these and other interactions, *Why Religion Matters*,[1] is in the best tradition of William James, Martin Buber and Marcus Borg: a skillfully written and credible account of the relevance of religious perspectives in modern life.

As a central metaphor for the first half of his book, the author introduces his readers to a *tunnel*. Modern Western culture has, in Smith's view, left the entrance of the Platonic cave — where shadows of reality flicker occasionally on the walls — to crawl deep underground, leaving behind 'the religious dimension of human life' which was a central element of the traditional world picture. The floor of Smith's tunnel is provided by *scientism*, a dominating modern view that is carefully distinguished from *science*. Whereas science — defined as an openness to evidence — is rated 'on balance good', scientism adds

> two corollaries: first, that the scientific method is, if not the *only* reliable method of getting at truth, then at least the most reliable method; and second, that the things that science deals with — material entities — are the most fundamental things that exist.

A central aim of this book is to challenge the claims of scientism, thereby helping our culture evolve toward a more balanced perspective on the nature of human existence. The ageless questions (Where do we come from? Who are we? Where are we going?), it is persuasively argued, do not become meaningless merely because their answers are not found in someone's theoretical formulation.

To emphasize the cultural dimensions of scientism, Smith develops his metaphor as follows. The tunnel's left wall is assigned to the universities, where academic success all but requires formal deference of fledgling faculty members to the materialistic and reductive perspectives of scientism. 'The modern university is not agnostic toward religion; it is actively hostile to it,' he asserts, and the

---

[1] **Huston Smith**, *Why Religion Matters: The Fate of the Human Spirit in an Age of Disbelief*, HarperSanFrancisco, 2001, IBSN 0-06-067099-1 (cloth).

humanities have been elbowed into a marginal status. Comprising the right wall is our legal system, which 'rightly assumes that theism is a religious position, while wrongly assuming that atheism is not.' Finally, the roof of the tunnel represents the media, which, as Smith amusingly shows, profoundly misrepresented the background and events of the 1925 Scopes trial in Dayton, Tennessee. Ever ready for a good fight, the Fourth Estate tends to encourage extreme views, giving both religious fanatics and reductive materialists public notice far beyond their numbers.

At this point, I imagine some (many? most?) *JCS* readers may be shaking their heads, preparing to set aside yet another religious rant that would return us to dark days before the dawn of science, but Smith sees our cultural history in broader perspectives. Different eras offer cores of wisdom that remain valuable to all: religion from traditional times and science from the modern era (reflecting Henry Adams's metaphors of 'the Virgin and the Dynamo' [Adams, 1996]), and finally the sensitivity to social justice that has emerged in the postmodern world. All such human experience can and should contribute to the development of Western culture in the present century. For those who insist that loyalties to both religion and science are incompatible, one can point to such eminent researchers as William James, Charles Scott Sherrington, Erwin Schrödinger, and John Eccles, who have persuasively argued otherwise. Beyond faith, what is the basis for their views?

According to Schrödinger (1983), a key element in the development of a world view is a *sense of wonder,* which scientism sadly lacks. Thus:

> The man who has never at any time felt consciously struck by the extreme strangeness and oddity of the situation in which we are involved, we know not how, is a man with no affinity for philosophy — and has, by the way, little cause to worry. The unphilosophical and philosophical attitudes can be very sharply distinguished (with scarcely any intermediate forms) by the fact that that the first accepts everything that happens as regards its general form, and finds occasion for surprise only in that special content by which something that happens *here today* differs from what happened *there yesterday*; whereas for the second, it is precisely the *common* features of all experience, such as characterise everything we encounter, which are the primary and most profound occasion for astonishment; indeed, one might almost say that it is *the fact that anything is experienced and encountered at all.*

Might not a sense of wonder — in addition to the invention of tools and language, and the exercise of logic — help to distinguish humankind from our less gifted biological relatives?

Stated briefly, the issue between science and religion is this: some believe that the greater has emerged from the lesser in the course of biological evolution, whereas others hold that something called *Spirit* is fundamental, from which elements of our presently perceived reality are derived. That this question is relevant to consciousness studies is emphasized by Smith, who states:

> If consciousness is not simply an emergent property of life, as science assumes, but is instead the initial glimpse we have of Spirit, we ought to stop wasting our time

trying to explain how it derives from matter and turn our attention to consciousness itself.

On both sides of the question, it should be noted, we are dealing with *beliefs* rather than established facts. As a working hypothesis, scientists assume that the greater can be understood in terms of the lesser; that is their role, and they play it well, becoming ever more convinced of the rightness of their view with each new success. The 'religiously musical', on the other hand, take some sort of Supreme Intelligence as the ultimate cause, seeking to understand human nature in the context of traditional religious experience. Adherents of both perspectives must admit that the issue is open, for there is much about reality that we shall never know. But what it is that we *do* know?

There is little doubt, first of all, about the practical value of religion. A central pattern in virtually all cultures, belief in some benign presence — be it polytheistic, monotheistic or mystical — provides individuals with a sense of place, of belonging in the universe, bringing many beneficial psychic and physiological side effects. At the group level, additionally, the social fabric among believers becomes more strongly woven and sewn together, with all the advantages that implies for a well functioning community. Some take these facts as positive evidence for the existence of Spirit. For others, the same observations provide reasons for treating religious beliefs with extra caution, aiming to avoid confounding comfort with conviction.

Beyond such psychological and cultural observations, there are many examples of *transcendent experience,* in which an individual feels a sudden and unexpected sense of unity with the cosmos, a flood of emotion laden awareness far beyond the levels of ordinary perception. Among several classic examples is Buber's *'Ich und du'* perception of a linden tree (Buber, 1970), but the phenomenon is not at all restricted to religious adepts and mystics. Transcendent experience visits people of all sorts, beliefs and backgrounds, including a significant number of scientists as the internet archives of Charles Tart (no date) have recently shown. One reason that the ubiquity of transcendent experience is not more widely recognized in our culture may be that we tend to view them as pathological, ignoring the deep pleasure they often bring. (Indeed, William James himself assigned such a very strong personal experience to a 'story told to him by a friend' [James, 1997].) Are transcendent experiences to be taken as examples of the cosmic intelligence seeping downward into benighted souls? Or, as has been suggested by Abraham Maslow (1968), might they be a normal human activity, surging upward from time to time in healthy minds?

Another relevant question, discussed in some detail in this book, is the nature of reality. Although the reductive scientist would restrict this term to the dynamics of our constituent molecules, this is not a necessary position, even for a physicalist. (Those who would limit reality to chemical substances should consider the unicorn. Although not part of biology, this creature exists in the mind and in literature, realms that Karl Popper has respectively labelled 'World 2' and 'World 3' [Popper and Eccles, 1977].) Cultural patterns, which have tenuous physical bases, can be important determinants of human experience, and it seems to many students of consciousness that the mind's dynamics are largely independent of their biological bases.

Above the material aspects of the universe are arranged several qualitatively different levels of reality, which Smith finds in all of the major religions (Christian, Muslim, Hinduism, Buddhism, and the Chinese Religious Complex) to take the general form

Spirit

soul

psyche

body

with each higher level encompassing the lower in multiple degrees of reality. Again, the difference between the religious and scientific perspectives is this: under the religious view, it is supposed that the lower levels *derive from* Spirit, whereas science assumes a hierarchical structure of the form

soul

psyche

body

material

with the upper levels *emerging from* the material.

Interestingly, the term 'emergence' does not appear in Smith's index, an oversight suggesting that even he may be overly influenced by the reductive claims of scientism.

In my opinion, Smith goes a bit too far in his criticisms of Darwinism. While rightly insisting that critics of the his perspectives set aside the dubious claims of fundamentalists and focus their attentions on the best of modern religious thought, he should not base objections to current evolutionary theory on the simple minded assertions of certain genetic reductionists. Punctuated evolution, for example, is not some vague exercise in hand waving, trotted out to evade the fossil evidence; this phenomenon is very typical of interacting nonlinear systems, which a collection of interacting species quite clearly is. Furthermore, no mention is made of important recent works by biologists like Stuart Kauffman (1996) and Brian Goodwin (1994), showing how intricate and interesting forms of life may emerge quite unexpectedly from seemingly simple underlying dynamics. Although there remain many open questions in evolutionary theory — not the least of which is how life managed to emerge at all from the boiling Hadean seas, some three to four billion years ago (Eigen and Schuster, 1979) — present formulations are not obviously wrong and current progress is encouraging, to the scientific community at least.

Smith also incorrectly claims that the scientific worldview excludes 'topdown causation'. Although this assertion may be the *opinion* of certain scientists, it is by no means a conclusion of modern science. The assumptions that present day science excludes are twofold: (i) Aristotle's *causa finalis* (represented by Spirit in the above hierarchical diagram), and (ii) the downward action of his *efficient cause*. Put in a different way, science currently excludes the possibility that mental phenomena can alter the laws governing chemical interactions in the brain's constituent neurons. (The forces acting between a certain carbon and oxygen atom in a protein of my brain do not depend upon what I happen to be thinking.)

On the other hand, downward actions of Aristotle's *material cause* and *formal cause* are certainly present in the brain. The emotion of fear, for example, stimulates the release of adrenaline into the bloodstream, which in turn influences the dynamics of neural behaviour: a clear example of the downward action of a material cause. The phenomenon of learning — which none can doubt — provides a ready example of the downward action of Aristotle's formal cause. In the wake of a psychologist's training schedule, some physical aspects of neural structure are altered (perhaps dendritic spines?), thereby changing the boundary conditions (a formal cause) influencing subsequent neural dynamics. Such examples of downward causation are of great interest and relevance to current research in both neuroscience and studies of mind (Andersen *et al.*, 2000).

Finally, those who suppose that a benign intelligence guides the universe must say something about the problem of evil. If a kindly God is all in all, as some mystics tell us, it follows that the appearance of evil is merely an illusion: a view that is difficult to square with certain events in the century that we have just left behind. In response to this problem, Smith offers a suggestion.

> If a two-year-old drops her ice-cream cone, that tragedy is the end of the world for her. Her mother knows that this is not the case. Can there be an understanding of life so staggering in its immensity that, in comparison to it, even gulags and the Holocaust seem like dropped ice-cream cones?

To me, this seems a stretch. I feel closer to Martin Buber, who insists that we must accept the reality of evil as an empirical fact, wherever this might lead our theology (Buber, 1980).

But if (as one of the grubby Tech students that he was formerly charged to humanize) I find Smith's arguments for a benign intelligence unconvincing, so what? Perhaps like Marcus Borg (1995) I am still overreacting to fanciful tales recalled from confirmation class. Perhaps I have spent too much time digging in the tunnel of modern science to become religiously musical. Whatever the reasons, I would not venture to advance my merely personal inclinations as scientific conclusions.

In many ways, indeed, Smith's views fit quite comfortably into my overall sense of things. Sharing his distaste for scientism, I am regularly amused by the fatuous claims for the philosophical importance of trivia advanced by assertive scientists in the *New York Times, Scientific American, Discover,* and so on. Reality is intricate far, far beyond the reaches of our collective imagination, and the sooner we scientists choose to recognize this evident fact, the better for all. In a final chapter of *Why Religion Matters* entitled 'We could be siblings yet' this view is underscored in the following terms:

> The religious sense is visited by a desperate, at times frightening, realization of the distance between [fundamental] questions and their answers. As the urgency of the questions increases, we see with alarming finality that our finitude precludes all possibility of our answering them.

What fair minded person could disagree? Smith's generous and eloquent call for a sense of joint enterprise in exploring the mysterious dimensions of human experience should be read by open minded representatives of both science and religion and accepted by all who share Schrödinger's sense of wonder.

References

Adams, H. (1996), *The Education of Henry Adams* (New York: Modern Library).
Andersen, P.B., Emmeche, C., Finnemann, N.O. and Christiansen, P.V. (2000), *Downward Causation: Minds, Bodies and Matter* (Denmark: Aarhus University Press).
Borg, M. (1995), *Meeting Jesus Again for the First Time: The Historical Jesus & the Heart of Contemporary Faith* (San Francisco, CA: Harper).
Buber, M. (1970), *I and Thou* (New York: Charles Scribner's Sons).
Buber, M. (1980), *Good and Evil* (New York: Prentice Hall).
Eigen, M. and Schuster, P. (1979), *The Hypercycle: A Principle of Natural Self-Organization* (Berlin: Springer-Verlag).
James, W. (1997), *The Varieties of Religious Experience* (New York: Macmillan).
Goodwin, B. (1994), *How the Leopard Changed Its Spots: The Evolution of Complexity* (London: Charles Scribner's Sons).
Kauffman, S. (1996), *At Home in the Universe : The Search for Laws of Self-Organization and Complexity* (New York: Oxford University Press).
Maslow, A. (1968), *Toward a Psychology of Being* (New York: Van Nostrand).
Popper, K. and Eccles, J.C. (1977), *The Self and Its Brain* (Berlin: Springer-Verlag).
Schrödinger, E. (1983), *My View of the World* (Woodbridge, CT: Ox Bow Press).
Tart, C. (no date), *The Archives of Scientists' Transcendent Experiences* (TASTE), (http://issc-taste.org/index.shtml).